Weeping Sandstone
The Geology of the Great Kanawha Valley

Bob Kessler

PublishAmerica
Baltimore

© 2003 by Bob Kessler.
All rights reserved. No part of this book may be reproduced in any form without written permission from the publishers, except by a reviewer who may quote brief passages in a review to be printed in a newspaper or magazine.

First printing

ISBN: 1-59286-173-3
PUBLISHED BY PUBLISHAMERICA BOOK PUBLISHERS
www.publishamerica.com
Baltimore

Printed in the United States of America

Dedicated to my wife,
Judy

Acknowledgments

I wish to extend my appreciation to my oldest son, John, for contributing to this project over the years in many ways but especially John's invaluable help manipulating computer programs, files and all those other computer things necessary to put a book like this together. To my daughter, Jennifer, who accompanied me on many field trips and waited patiently as I continually stopped and hammered away at any rock I came across. Jennifer could spot a piece of petrified wood in a rubble-choked streambed from twenty feet. To my youngest son, Joe, who shared my enthusiasm for the outdoors by accompanying me into the field on countless trips; rain, shine, and snow. Joe probably knows as much about the geology of Kanawha County and petrified wood as anyone. And, of course, to my wife Judy — to whom this book is dedicated. Without her tolerance for dirt, boxes, loose rocks, and fossils scattered about the house, and without her unwavering help and support all these years I would not have, or could not have, written this book.

I want to thank and acknowledge the efforts of Ray Lewis who first found the undisputed location of in-place petrified logs and who directed me to many other sites of interests, some of which ultimately ended up as pictures in this book.

CONTENTS

Introduction .... Page 11

Chapter One *Where Are We* .... Page 18
    Cenozoic Era .... Page 25
    Mesozoic Era .... Page 27
    Paleozoic Era .... Page 31

Chapter Two *Stone Story* .... Page 52
    Some Stone Stuff .... Page 56
    Sandstone .... Page 57
    Siltstone .... Page 62
    Shale .... Page 63
    Clay (Claystone) and Mudstone .... Page 65
    Limestone .... Page 67
    King Coal .... Page 69
    The Pennsylvanian Period .... Page 71

Chapter Three *Your Backyard Rocks* .... Page 77
    The River .... Page 86
    The Parts .... Page 89
    Bed Time .... Page 90
    Monongahela Series .... Page 95
    Conemaugh Series .... Page 98
    Allegheny Series .... Page 108
    Pottsville .... Page 116

Chapter Four *Fossil Trees of the Kanawha Valley* ... Page 128
    Modes of Fossil Preservation .... Page 130
    Lycopods (Scale Trees) .... Page 140
    Calamites .... Page 146
    Cordaites (The "Wood" Tree) .... Page 150

Chapter Five *Fern, Vines, and Fern-Like Fossils* .... Page 159
        Ferns .... Page 159
        Fern-Like Seed Plants .... Page 164
        Fossil Animals .... Page 174

Chapter Six *Our Heritage* .... Page 186

Epilogue .... Page 207

The Great Kanawha River Valley

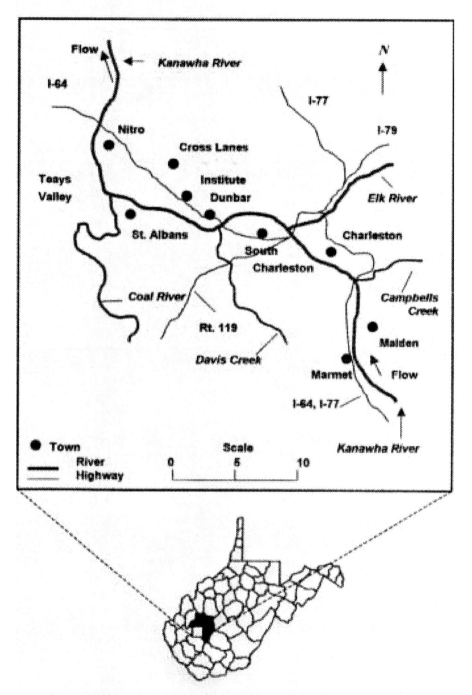

Kanawha County (in black), West Virginia

INTRODUCTION

"The name Kanawha was probably derived by evolution from the name of a tribe of Indians (a branch of the Nanticokes), who dwelt along the Potomac, and westward to New River. They were variously called or spelled by different authors at different times, Conoys, Cousise, Canawere, Colmawes, Canaways and Kanawhas. The spelling of the name has been very various, in addition to the ways mentioned above, including nearly all practical methods commencing with C or K. Wymans map of the British Empire in 1776, calls it the 'Great Conoway or Woods River,' The act of the Legislature of 1789 forming the County spelled it Kenawha. In an original report of survey made by Daniel Boone at the mouth of the river in 1791, and now in the writers possession, he spelled it 'Conhawah.' The accepted spelling now is Kanawha, probably never to be changed again." (Dr. John P. Hale, *Trans-Allegheny Pioneers*, 1886.)

Before 1760 the Kanawha River, as well as the New River, was known as Wood's River, named for Colonel Abraham Wood, British commander of Fort Henry, located on a bluff near the falls of the Appomattox River and close to the present site of Petersburg, Virginia. A merchant and land trader, Colonel Wood set out in 1654 to explore western Virginia beyond the Blue Ridge Mountains, where he discovered the great river near what is today Radford, Virginia. One hundred years later, Mrs. Mary Ingles and several others who had settled in the vicinity of Wood's River, in almost the exact spot first seen by Colonel Woods, were taken captive by embittered Shawnee. Forced by their captors down river to the Ohio, these beleaguered settlers reportedly are the first non-natives to see the Kanawha Valley. As a matter of history, Mrs. Ingles saw it twice in a four month period, once going down river in the company of the Shawnee and once hightailing it back upriver while making her escape in the company of a troubled, old Dutch women.

Although this seems a long time ago, human occupation in the Great Kanawha River Valley extends into the past some twelve thousand years, a time when huge ice sheets had just enjoyed their last hurrah and were stubbornly retreating northward across Ohio and back into Canada from whence they came. The weather was changing; after a hundred thousand years in the shadow of a mile high ice sheet, the relatively new Ohio Valley and its tributary valleys finally began to warm. Cool weather plants like the shrub of the blueleaf birch, aspen, balsam poplar, and northern white-cedar slowly gave way to the great American chestnut, green ash, yellow buckeye, oak, and sweetgum. Into this changing wonderland the Native Americans ventured.

Hundreds of books and articles have been written on the history of the Kanawha River Valley which, when counting the early Native Americans, covers the last twelve thousand years or so, although most have been written on human events that took place only in the last three to four hundred years. Prior to this there was little to no documented record keeping by the native people who lived, hunted and/or just passed through the valley. What we do have, though, or at least we used to have, is evidence of early Americans in the form of their exceptional craftsmanship; burial mounds and living structures which used to line the floodplain up and down the river, and an array of implements used in their everyday life from potsherds to spear points are still found in backyard gardens today.

There were dozens of mounds in the Charleston area alone, also in South Charleston, and toward the mouth of Tyler Creek; Aarons Creek, Dutch Hollow, Finney Branch, and Institute — all so close together that if you walked down the valley you would always have the next one in sight. Many homes in the valley today, unbeknown to their current inhabitants, are setting quietly on long disregarded and permanently obliterated "Indian" burial mounds. It seems our famous forefathers and original land "owners" (most of our streets are named after them) had other things in mind than contemplating antiquity. Although they did, for whatever reason, spare one in what is today South Charleston. But is it not offending, if not downright degrading, to build steps on it and a summit platform so we can all get a better view of the aging chemical factories.

History describes, as best it can, past human events, activities, and discarded remains; that is what history is all about, but prior to three or four hundred years ago human history in this region was scant. Somewhere in this duration (from now to about twelve thousand years ago), the science of archaeology disconnects with so-called modern history and then extends back as far as human evidence allows it; in other words, no humans, no archaeology, no history.

This book is also a chronology of the Kanawha Valley, but it is a saga of another color, there are no people in this book (alas, it is not a history book) but many pioneers nonetheless; no good or bad endings, no plot, and no specific heros — it is just the way it was. The real drama occurred in the Kanawha Valley before history as we know it, a time referred to as prehistory (or prehistoric); that is what this book is about. A time long before human input and political endeavors, a time before fur trappers, land grabbers, timber and coal barons, and railroad builders, even long before the first Americans crossed the Bering land bridge from Asia into Alaska, a time so far away it is incomprehensible.

Prehistoric events were not written down by some medieval scribe on scrolls of papyrus, nor were they recorded in the hieroglyphs of ancient Egypt; prehistoric events were instead written down by Nature in the thickest book ever created — the earth itself. Weather reports are there, catastrophes are there, stories of plant and animal communities are there, and death is there. All one has to do is learn the language and the stories are endless. Nature has two faces, one of constancy and one of randomization. Throw a rock into the air and it will always fall to the ground, throw it again a thousand times over and it will never come to rest in exactly the same place as the first. It makes you wonder how you ever did it the first time.

It does not take a keen eye to notice the trees and rocks of the Great

Kanawha Valley because they are everywhere. Most of us have a natural curiosity about trees; we like to know their names and be able to identify them. Maybe it's because trees have such colorful and appealing names; names like slippery elm, honey locust, persimmon, and weeping willow (and some others I mentioned above). Rocks, on the other hand, have names like shale, sandstone, limestone, and clay. Not as attractive as tree names for sure but the rocks are just as remarkable in their beauty and uniqueness. Think how enjoyable it would be when in the forest (or in your own back yard) you could identify the common trees and you knew the rock types as well as their "history." A hike would become something more than just a hike; it would become an excursion of discovery. And who knows, you may be lucky enough to spot a majestic and massive outcrop of the beautiful, mottled gray to blue honey limestone overlain by the shear cliffs of the rare and endangered, blacked-streaked weeping sandstone.

These may be two entirely different subjects, plants belonging to the science of botany, and rocks belonging to the science of geology, but when combined they make up the greater part of what we see. They are the scene, or backdrop, where all our activities are played out. Even though we may or may not think about them much, they are a part of us, they are a part of our environment and we are a part of their environment. Think about it, especially when hiking, that is about all there is, plants and rocks, with an occasional bird or chipmunk here and there, and usually plenty of insects. So why is it so much fun to get out in the forest if that is all there is? It's fun because we humans have a long evolutionary heritage in the forest; we feel it in our bones, the solitude makes us uneasy and at the same time excites and comforts us. There is no other place that can incite these seemingly contrasting feelings at the same time then when we are in the forest, alone with the plants and the rocks.

I want to stress one thing first — to just memorize the name of objects is only half the fun, most anyone with a obliging memory can do that. When it comes to Nature, the whole point of knowledge is to be able to really deep down appreciate the wonder of it all. You have to get out and look around. Get out in the woods on a blustery winter day when all your friends say it's too cold; climb, hike, and take pictures. Take a lunch, or just some beef jerky, sit beside a cold, half-frozen creek or under a rock overhang draped with icicles. Look around, enjoy, and if possible always try to be with someone with equal enthusiasm and scientific curiosity.

Starting a plant or rock collection (or both) is a great way to learn the common names of all the local types but more importantly it is a great way to begin to appreciate the beauty of Nature that is so often overlooked. Plants are the product of millions of years of evolution; the rocks, especially here in the Kanawha Valley, were formed so long ago it is unimaginable, and that is true. It is difficult enough to even imagine the time when our grandparents grew up, let alone when our great (now say "great" 12 million more times) grandparents grew up. A long time indeed. You and I, and every other living organism (plant and animal) now on earth, are the product of an unbroken line of ancestors going back to the beginning. Had the line been broken just once, you, or I, would not be here. A sobering thought indeed. Getting back to the grandparents for a second and just for the sake of it, if we consider twenty-five years per generation then 12 million "greats" multiplied by twenty-five years per generation equals 300 million years (12,000,000 x 25 = 300,000,000). That is about how old the rocks are in the Kanawha Valley, give or take a million. This particular time in prehistory is called the Pennsylvanian Period. Our ancestors looked different back then I bet.

To learn the names of the common trees found in the Kanawha Valley I recommend a good tree identification guidebook for starters. But even with a book you have to get outside where the trees are; take your book with you and study and compare the leaves. This will get you started, but if you are like most, the more you learn the more you will want to know and the more enjoyable it will become. To learn something about the rocks of the Kanawha Valley in everyday language, I recommend this book (because it is the only one there is) which is intended for readers of all ages who have an interest in rocks (and *who* doesn't?) or who are maybe just curious about it all. This book is about the rocks of the Kanawha Valley, how they were formed, what type rocks are found here, their age, and what can be found in them.

Rocks are everywhere; we build on them, we skip them across a pond, we grind them up and make concrete and asphalt soup out of them — we even eat them (salt). If they get in our way, we blow them up and dump them somewhere else. Do rocks get any respect at all? What exactly are rocks and where did they come from in the first place? What the heck are they made of and how did they pile up so high in the Kanawha Valley?

When I first saw the pictures taken by NASA's Pathfinder of the Mars' landscape, I focused on one particular small rock lying there in the red dust among all the others. It looked like it had been lying there a long time, but nothing gets anywhere without something putting it there. What force put that rock right there on that spot way up on Mars? Maybe a volcano blew up and scattered rocks all over the place, or maybe 50 million years ago a comet slammed into Mars and knocked that little rock right there. I wondered what was the point anyway, if no one was around to notice it? Was it put there with no regard whatsoever to the human race that would someday inhabit Earth? My answer to the first question is, there was no point — it was a random event however it got there; it just happened. My answer to the second question is, yes.

Although we may never know, in my lifetime at least, that much about the rocks on Mars, we do know where the rocks came from around the Kanawha Valley. We have defined what kind they are, how old they are, and in what type of environment they formed. The study of rocks falls within the science of geology, which in so many words is defined as the science that deals with the "history" of the earth. This book has been written to provide the reader with an introduction to the geology of the Kanawha Valley, principally the area between, say, Marmet on the south side of the river and Malden on the north side, and west to the Saint Albans-Nitro area, with some liberties north and south of the river. I will discuss the different rock types present, what they look like, fossils, petrified wood, flint, and much more. But you don't have to live in this specific area to enjoy and learn about rocks, with the exception of the second half of Chapter Three where I mention certain outcrop locations and cities in the valley; the information discussed herein is applicable to most of the eastern half of West Virginia where Pennsylvania-age rocks dominate. I have attempted to include topics of general interest which are applicable to all sedimentary environments whether in the Kanawha Valley, along the Big Sandy, Hughes River, or Wheeling Creek.

A book like this, like any present-day science book, is fashioned on the backs of earlier scientists, in this case those field-hardened and committed geologists of the early West Virginia Geologic Survey. Much of the information I worked with in this book, especially the information provided in Chapter Three, was taken from the (then) West Virginia Geologic Survey's publication *The Geology of Kanawha County* (C.E. Krebs & D.D. Teets, 1914).

I have spent the last thirty-five years in the company of my sons, John and Joe, and my daughter, Jennifer, climbing, describing, and collecting fossils from the hills that define the Kanawha Valley; they conquered the seven-hundred-foot slope of old Bald Mountain when only six, six, and five respectively and gazed with young eyes on the valley below.

To see rocks and grasp their physical presence as part of the landscape, to actually touch them and understand their origin, and most of all to appreciate their antiquity, you must go where the rocks are. Get out in the forest, get off the path if you can and hike up a creek bed. Look down, wade in the water — where did all those rocks and boulders come from anyway and what kind of rocks are they? Let the sun shine, let it rain, who cares, it is a beautiful day to be outside. When you finish this book you will not only know where you are geographically, but you will know where you are geologically as well.

Throughout the book I tried to be as accurate as possible with regard to specific rock formations (or rock beds) names and their placement as they occur in the valley. But, because geology is not an exact science, like say chemistry, there is room for error. What I have provided was developed from my own field and research experience while following the basic terminology and rock formation chronology established in the West Virginia Geological and Economic Survey's 1914 publication mentioned above. So, it's fair to say, any errors are mine.

Chapter One
Where Are We?

Time gone by, I wonder why.
Things will never be the same,
what a shame.

Since this chapter is about time I was actually tempted to begin with the timeworn, "Once upon a time ...etc., etc.", but the more I thought about it I realized that in this case, when discussing the age of the Kanawha Valley rocks, time really has no practical meaning. Who can comprehend even a million years ago? I can't; how am I to expect anyone else to? How am I to expect anyone to think back beyond civilization, beyond the many advances and retreats of the glaciers, beyond the great age of the mammals and even far beyond the dinosaurs? How can I expect anyone to think back to a time when the largest land animal on earth was a most obnoxious, slimy-mouthed amphibian? I can't comprehend a million years either, let alone 300 million years but I have come to terms with it the same way I have come to terms with the real time of my great-great grandfather...I know he had to exist in his time because I exist in mine.

At least for a while, I am going to talk about time and where the rocks of the Kanawha Valley "fit" into the earth's overall geologic time. The earth's geologic time, looking backwards, goes from now back to the earth's beginning and is associated with the rocks created during successive rock building time periods. The resultant chart is called the geologic record (Table 1, page 24). So, the geologic record (or chart) just means a listing of the identified and accepted age of the earth's rocks from the youngest known to the oldest known.

One of the fundamental doctrines of geology, as elementary as it seems, is that younger rocks are always above older rocks, unless some great force pushed them upside down. Looking at a road cut it makes sense that the rocks toward the bottom of the cut are older then the rocks toward the top. And this is true because the rocks that are covered with other rocks had to be there first. So, the deeper you dig, so to speak, the older the rocks you encounter.

The rocks of the earth's crust have been carefully dated, studied beyond belief, and assembled into the table on Page 24 which lists the rocks from the youngest to the oldest. The sequence of rocks in most tables usually has the youngest rocks at the top and the oldest rocks at the bottom, just like if you dig down into a very deep hole identifying and dating the different rocks as you go deeper; although, just for the record, no single place on earth has all the rocks of the geologic table represented — but some parts of England come close. This is what is meant by the geologic record, a visual record of the sequence of all known rocks from the youngest to the oldest, even though you can't find them all in one deep hole.

I briefly mentioned time in the introduction and said the age of the rocks of the Kanawha Valley was 300 million years old, and then I qualified it by saying, "give or take a million." The reason I qualified the statement was simply because it is impossible to exactly date rocks so far back in time, even with today's sophisticated dating techniques, to within a million years or so. But in most cases when you are talking about events of this magnitude, within a million years or so seems close enough. Are the rocks of the Kanawha Valley too old to find dinosaur fossils in them, are they too young, or are they just right?

Time is a curious thing; when I'm in an audience listening to some boring speaker babble on about a subject of which I have little interest, time goes rather slow, at least I think it does. When I'm out in a newly-plowed field on a breezy summer's day with my son Joe, methodically scanning each new furrow for flint chips or arrowheads, time goes fast. Is it just a matter of interest, or maybe awareness? Did the 300 million years since the rocks of the Kanawha Valley were laid down go fast or slow? Since I have no awareness of it, it seems to me it went pretty fast; I could say in an instant — wherever I was I must have been enjoying myself. That's the trouble with time. Simply because I don't remember the last 300 million years it didn't really happen as far as I am concerned; time started when I started and will end when I end, all the rest just makes good reading. But, fortunately for us it did happen, minute by minute; for some obscure reason time doesn't need me and it will probably just keep ticking away when I'm gone.

Time may be thought of as the interval between two events, which seems logical because if two events happen at the same time (same exact instant) there is no time between them. Using this logic, humans have identified "time" intervals and have fashioned ways to compare them. Some genius a long time ago more than likely observed that every once in a while the sun was in the exact same place in the sky as it was once before (event #1 and event #2). Not only being smart but patient as well, he, or she, counted (or placed pebbles in some cave somewhere) the number of sunrises between these two events, and as it turned out when the second event rolled around there were 365 pebbles in the pile. Even if this genius didn't know how to count, there were still 365 pebbles in the pile. Subsequently, the interval (measured by the number of pebbles) between the two sun events was called a "year" or some such utterance that meant the same thing, and the collection of smaller intervals, in this case identified by the pile of pebbles, were called "days" — hence a pile of days equals two suns and a smaller pile of days meant two moons, and so forth.

This may seem over simplified but someone identified these intervals and gave them names so they could use time to their own advantage. With this kind of high-tech knowledge, our ancient ancestors must have learned the right pebble (day) to move north to meet the migrating antelope herd and then move back south before the winter set in. But, like it or not, the intervals were there before we gave them names. We obviously didn't invent time, humans were just the first life form to really notice it beyond a day and then quantify it.

So, what does all this have to do with the geology of the Kanawha Valley? I think it is interesting to stress that time was going on before anyone figured out how to use it; events were taking place long before the human species was around to witness them. There existed before our species an almost endless number of past sunrises and no one was counting. The sun came up, stayed up a while, and then went back down, day in and day out. The nights were long and dark — really dark. It rained, it snowed, trees grew to old age, died and fell down to make room for another. Rivers ran one way for a few million years and then ran in some other direction or just dried up. Big bugs ate little bugs and then something ate the big bugs. All things happened that could happen. And it all took place without us.

For millions of years the earth was bursting with untouched beaches, endless virgin forests; cool, clear streams, clean rivers full of fish and other water life, and wild animals galore; this was no mere garden, it was Eden on a global scale, and it lasted forever. The question is, why? In human terms, this seems like a vast amount of time and beaches wasted; there was no one around to really enjoy it — a developer's paradise with no developers. Nature was literally on her own for an awfully long time, and in Nature, what works, works well, and what doesn't work, gets left behind. Nature is not the least bit vengeful as we have been led to believe. Nature has no shame and chance is the name of the game.

Even though we were not around to see prehistoric events in person, we can still read about them. Each rock sequence, or rock layer, may be thought of as a page in a book. As you turn the page, the further back in time you go. We can read of life and death events, we can read about where rivers once flowed, about catastrophic floods, tropical forests, swamps and marshes, and sand dunes. Written in the rocks are environmental changes, sometimes abruptly, sometimes gradually, where once swamp conditions existed, sluggish river deposits suddenly appeared. We can see it all because each distinctive rock type or fossil tells its own story just by being what it is.

Take for example the small, clam-like sea creature called the brachiopod (Figure 5). We know that all brachiopods that ever lived, lived in the sea. There were no freshwater brachiopods and to this day they still live in the sea. So, if a brachiopod fossil is found in some shale layer in a road cut or quarry somewhere, you know immediately the sea must have once covered the spot in which the fossil was found, and the shale layer itself must have been deposited in a relatively-shallow, protected area of the sea away from any influence of rivers. Rivers are notorious sand and silt carriers and shale is an accumulation of mud. The brachiopods have been studied, dated, and named so when one is found and identified the relative age of the rock (in this case the shale layer) can be determined, since the brachiopod had to live at the same time it was covered with mud. Amazing!

Most books written about geology usually has some metaphoric example to illustrate how much time has passed when compared to how long ago we humans migrated out of Africa. These examples usually go something like this: If you relate the time from the earth's beginning until now to one year then humans have only been around the last few minutes, or, if you compare the age of the earth with a mile of highway, the first human appeared only a couple inches from the end. These examples sound good and do illustrate somewhat the time distances between "us" and earth's ancient beginning, but even so (as I mentioned before) I find it futile to comprehend any concept of time this far back, regardless of the examples and good intentions of writers. Time just has no handles, it's as simple as that. Yet, I believe most people do accept quite well the fact that the earth is incredibly old and that it has survived many changes.

Just how old is the earth? Present scientific wisdom dates the earth to around four to four and a half billion years old. Since dating techniques are beyond the scope of this book, all ages referenced reflect current recognized and accepted dates.

Table I is a representation of the geologic time scale from "us" back to the current accepted age of the earth. This table has been divided into Eras (far left), Periods (next right), approximate number of years to the start of a particular period, and a couple of columns showing when some important animals and plants first showed up in the fossil record. From time to time (especially in this chapter) reference will be made to this table to keep the different geological events that are being discussed in proper perspective. The sequence of ages shown in the table is very important to understanding the earth's geological history and where the rocks of the Kanawha Valley belong.

What you see in the table is the fundamental and identifying catalogue of the earth's history used by the scientific community. This is the geologic record mentioned earlier. Different countries, and even different states, however, do apply in many cases different names for the various smaller or local rock formations which exist within the periods themselves. Sometimes this gets a bit confusing when studying the local geology of bordering states, but is of no concern here.

The geologic ages and rock formations contained within them will be referenced throughout the rest of this book, so it is necessary to examine Table 1 with some degree of attention — especially since the information in this table is the crux of what geology is all about. Another reason this whole table will be briefly reviewed is to provide the reader with a sense of the age of the rocks exposed in the Kanawha Valley relative to the rest of the ages identified in the table. So, if you will, bear with me on this for the next several pages while we proceed through the last 570 million years of earth's history. Believe me, it won't take near that long. Note first that the time preceding 570 million years ago will not be covered (see Table 1, Pre-Cambrian), this is so distant in time that it has absolutely no meaning with regard to our objective.

The first column in the Table (far left) lists the name of the Eras. Eras have been established on major earth-changing events, particularly major extinctions. There have been many extinctions throughout the 570 million years just mentioned but a couple of the greatest and most curious separate the Eras. There are three Eras to consider; the Cenozoic, the Mesozoic, and the Paleozoic. These three Eras make up 570 million years of geologic time, anything before this will just be referred to as the Pre-Cambrian. Each of the Eras are divided into Periods. In most cases when discussing the Eras the corresponding periods will be discussed; this is where the action is and some of the periods will, no doubt, be already familiar. Remember when looking at Table 1 that as you look down the page you are going further back in time; if you want to know how far back in time look at the third column with the heading "Years Ago." All the years referenced are, of course, approximate.

GEOLOGIC TIME SCALE

ERAS	PERIODS	YEARS AGO (X 1000)	FIRST APPEARED ANIMALS	PLANTS
Ceno-zoic	Quaternary	1,800	us mankind	
	Tertiary	65,000	dinosaur extinction	
Meso-zoic	Cretaceous	146,000		flowering plants
	Jurassic	208,000	birds	
	Triassic	245,000	dinosaurs small mammals (mass extinction)	
Paleo-zoic	Permian	286,000		
	Pennsylvanian	325,000	giant insects	coal forming swamps
	Mississippian	360,000	primitive reptiles amphibians & lobed-finned fish	"woody" trees
	Devonian	410,000	insects & spiders	seed & true ferns
	Silurian	445,000		land plants
	Ordovician	510,000		
	Cambrian	570,000	Jawless & boney fish brachiopods & trilobites	
Pre-Cambrian (Not to scale)		4,000,000		blue-green algae one-celled organisms

TABLE 1

Cenozoic Era

The first and most recent Era is the Cenozoic. The Cenozoic is divided into two periods, the Quaternary (that is where *"us"* is under the sub-column "Animals" in the Table) and the Tertiary. The Cenozoic began roughly 65 million years ago with the beginning of the Tertiary Period (see Table 1). A few words about the Quaternary and Tertiary Periods. The Quaternary is the period we are in right now, at least the very last part of it. The Quaternary was also the period that saw the great ice sheets (glaciers) that covered a large part of North America and Europe, the last of which receded only 10 thousand years ago. The ice ages occurred during the Quaternary Period in a time called the Pleistocene Epoch (not shown in Table 1). The glaciers did not extend into West Virginia; the farthest advance reached to a line from around Cincinnati going diagonally northeast through Ohio to around Youngstown.

When compared to the preceding periods, the Quaternary and Tertiary Periods saw an explosion of mammal life. Great mammoths (including the wooly mammoth) and mastodons roamed across North America. Mastodons, which were also giant elephant-like mammals, were more adept to forest living and most likely lived in the valleys and forests of West Virginia. One of the quickest ways to identify a mastodon from a mammoth (when all you have is the skeleton) is by their teeth. Mastodons had huge teeth with numerous sharp, upside down, cone-like projections on them which made the teeth ideal for forest browsing on twigs and branches. While mammoths also had huge teeth, they were basically flat with hard ridges running through them, ideal for grazing on the lush steppe vegetation.

There were several species of mammoths during the Cenozoic. The largest of all, the short-haired Columbian mammoth, reached a height of 13 feet to the top of his head and weighed an estimated 13 tons. If your house has 8 foot ceilings, look up and then imagine another 5 feet above that. The largest tusks ever found, an astonishing 16 feet long, belonged to a Columbian mammoth skeleton found in Texas. Although curved, if they were straightened, they would be twice as high as your ceiling. Incredible.

The woolly mammoth, which lived in the more northern and colder regions, was somewhat smaller at around 9 feet and weighed approximately 4 to 6 tons. Equipped with extremely long, coarse, reddish hair, small ears, and an ample supply of body fat, the woolly mammoth was very well adapted for a frigid life along the glacial margins. Tusks and bones of these huge creatures are still being dredged from gravel pits along the Ohio River. Great quantities of mammoth tusks are found each year in several frozen, northern countries like Siberia and sold by the locals. The majority of ivory sold today comes from mammoth tusks. Look closely the next time you see something carved from ivory, if it has a slight pinkish or orangish appearance with ghostly patterns of herringbone it is from the tusk of a mammoth.

Other large mammals that roamed these parts during the Quaternary and the latter part of the Tertiary include the saber-toothed cat (Figure 1), the flat-face bear, the giant ground sloth, Yesterday's camel, and the horse. Although the horse was originally native to the Americas, it died out rather suddenly several thousand years ago (as did most of the large mammals) and was reintroduced by the Spaniards in their quest for gold while annihilating the native peoples of South America. Actually, the horses, like the mammoths, were most likely hunted for meat and skins by early Americans, although I doubt if anyone hunted the saber-toothed cat.

The Quaternary and Tertiary Periods of the Cenozoic Era each have their own remarkable story and much has been written about each of them, but let's continue for now further back in time to about 65 million years ago to the beginning of the Tertiary Period where one of the most baffling mysteries of all times occurred -

Figure 1 — The saber-toothed cat was about the size of a modern-day African lion. With short, muscular legs and seven-inch fangs, the sabertooth hunted by ambush.

the extinction of the dinosaurs. If you look at Table 1, notice that the beginning of the Tertiary Period coincides with the end of the Cretaceous Period of the Mesozoic Era.

Mesozoic Era

The Mesozoic is divided into three periods, from the youngest to the oldest; the Cretaceous, Jurassic, and Triassic. The Jurassic, of course, was the period made famous by the movie *Jurassic Park*. If you didn't already know, now you know where it is in the scheme of things and how long ago the Jurassic Period really occurred.

The end of the Mesozoic Era, represented by the end of the Cretaceous Period (65 million years ago), saw the extinction of the dinosaurs and many other land and sea animals. Approximately 38 percent of all marine (sea) animal life went extinct and the percent of land animals was even higher, of which the dinosaurs were a significant part. The boundary between the Cenozoic and Mesozoic was thus drawn on this mass extinction, although scientists usually refer to this extinction boundary as the Cretaceous-Tertiary boundary, or the K-T boundary. The "K" was used instead of a "C" (for Cretaceous) to prevent confusion with another period yet to be discussed. Many of the small mammals lived through this extinction period and went on to become the dominate animal type in the following periods of the Cenozoic, evolving into the large mammals of the Tertiary and Quaternary Periods mentioned above.

So far scientists have been unable to agree on just what caused the mass extinction at the end of the Cretaceous Period; some say global cooling, falling sea-level, diseases, others think a comet or asteroid slammed into the earth 65 million years ago and did the damage. There is, however, evidence from the fossil record that the climate was slowly changing throughout the world toward the end of the Cretaceous Period and that there had been a gradual "die-off" of many animal species for several million years prior to the boundary between the Cretaceous and Tertiary Periods. This extinction boundary (the K-T boundary), is now identified in the real geologic column (in the actual rocks) as a thin layer of white clay containing what might be mineral evidence (iridium) of fallout from a catastrophic hit from a large comet. Below this line (below the clay layer that separates the Cretaceous and Tertiary Periods), there are dinosaurs and other common fossils of the day and above it many are gone, most notably the dinosaurs.

Was the extinction of the dinosaurs really caused by a catastrophic earth hit by a comet? Maybe, maybe not. Consider a thin, clay layer (an inch thick) several hundred feet up the side of a road cut like some of the high road cuts seen on the turnpike or Corridor G going south. This couple hundred feet of rock from the road up to the clay layer could represent several million years of sediment (rock) accumulation. Say that you could find numerous fossilized dinosaurs bones along the elevation of the road but halfway up there were less to be found. Three-quarters of the way up (toward the clay layer) there were even less to be found and by the time you get to the clay layer all dinosaur fossils were gone. Does this mean a comet (indicated by the suspect clay layer) rubbed them out or does it mean they were pretty much gone when it hit? Maybe no one has been lucky enough yet to find any beyond the clay layer.

The days of the dinosaurs were for the most part over at the end of the Cretaceous Period, with or without a comet. Mammals were becoming more numerous and they loved to eat dinosaur eggs. Major environmental changes were taking place causing lowering sea levels which dried up the inland seas and reduced the dinosaurs' habitat and range, and possibly diseases introduced from animal migrations from other continents when sea levels were low contributed to the dinosaurs', and a multitude of other animal species', demise. Everyone has an opinion but no one knows for sure — one thing is sure, at the beginning of the Tertiary Period the dinosaurs as we know them ceased to exist.

Whatever the cause of the dinosaurs' extinction, during the three Periods of the Mesozoic which lasted some 170 million years, they were one of the most successful animals groups that ever lived. There is no need to feel sorry for the dinosaurs because 170 million years is a long time, especially when you consider that the human species has only been around a couple million years. Do you suppose mankind will still be here in another 168 million years? Well, maybe, who knows.

Most scientists believe that if it were not for the extinction at the end of the Cretaceous, dinosaurs would still be the dominate animal today and the mammals would probably still be scurrying about the underbrush trying to escape being eaten or stepped upon. Dinosaurs ruled from the beginning of the Triassic, where they started out as rather small, half-mammal, half-lizard-like creatures, until their culmination 170 million years later represented by giant plant eaters like *Apatosaurus* (late Jurassic, Figure 2), and deadly meat-eaters like the infamous *Tyrannosaurus Rex* of the late Cretaceous Period. These two particular and rather famous dinosaurs — most children know their names and can identify them easily — actually never met. Probably a good thing for *Apatosaurus*. But there were other monsters for *Apatosaurus* to worry about during the Jurassic Period.

Figure 2 — One of the large, plant-eating dinosaurs of the Jurassic Period, *Apatosaurus* was 95 feet long (from the top of his head to the tip of his tail if you stretched him out flat) and weighed 20 to 30 tons. Although not as long necked and sleek of body as some of his cousins, it is believed *Apatosaurus* could rear up on his hind legs and reach 35 to 40 feet into the treetops for food; although, I doubt *Apatosaurus* had, or needed, this talent.

Tyrannosaurus Rex, which literally means tyrant reptile king, was the most dreaded and ferocious beast roaming the planet during the latter part of the Cretaceous. There have been many "King of Beasts" throughout geologic time, all seemingly more terrifying than their predecessor but the culmination of brutality was T-rex —a homicidal maniac with a mouth full of six-inch steak knives. An unscrupulous scavenger, T-rex would eat anything alive or gulp down with equal fervor anything already dead. T-Rex the Ripper.

The ancient ancestor of *Tyrannosaurus* and largest predator of the

Figure 3 —The great granddaddy of the T-rex, *Allosaurus*, was the curse of the Jurassic Period. Fast, agile and downright mean, *Allosaurus* had no rivals. *(Redrawn with permission from original drawing by Joe Tuccirone.)*

Jurassic Period was *Allosaurus* (Figure 3). *Allosaurus* looked a lot like T-Rex except not quite as big; however, *Allosaurus* had longer arms equipped with large claws for holding onto its food while eating, something T-Rex could not do as good but apparently did not need to. The devilish and quick-moving meat eater that chased the children in the movie *Jurassic Park* was the *Velociraptor,* a dinosaur that actually lived in the Cretaceous Period along with T-Rex. These were perilous times for plant eaters.

The first fossil evidence of birds is from the upper Jurassic Period. These "birds" had the skeletal characteristic and teeth of a reptile but were covered with perfect feathers and had wings. The most famous of which is *Archaeopteryx*, discovered in a limestone quarry in Bavaria, Germany, considered by most paleontologists to be the intermediate stage of evolution between a branch of land reptiles and birds. The true missing link.

The oldest known flowering plants are found in Cretaceous-age rocks of the Mesozoic Era. No one knows what color they were since colors don't fossilize but these ancient blossoms must have dressed up the landscape like nothing before — unless they started out green. The evolution of flowering plants was a giant step for plants as a whole since all flowering plants today can trace their ancestory back to these enterprising yet rudimentary plants. Walnuts, hickories, tomatoes, beans, roses, and lilacs to name a tiny few are all genetic beneficiaries of the Cretaceous flower, which, by the way, is supposed to have been the favorite food of the tank-like *Triceratops*, a Cretaceous-age vegetarian.

It seems there was a great arms race during the Mesozoic Era, as the carnivores (meat eaters) became more fierce and developed better weaponry to catch their prey. The herbivores (plant eaters) developed more efficient defensive mechanisms, like the *Triceratops'* horns or the giant size of the *Apatosaurus*. Some herbivores learned to run faster, others learned to fight in packs or developed skin camouflage and spikes. Sometimes the carnivores won and sometimes the herbivores won. Actually, the herbivores didn't really win anything, they just got to keep their skin and go on doing whatever they did. They could, however, lose. Each side became so good that it was not that easy for a predator to actually win a battle with an adult herbivore without risking its own injury. As a result, the predators usually went after the young or older, sickly animals. Whatever, each did what they had to do to survive, they knew no better because there was "no better." Each animal was born with certain physical attributes, the T-rex did not know it had six-inch razors in its mouth and looked and smelled nasty, and the *Triceratops* did not know he had three great spear-like horns and weighed five tons. What each did know was to react to the image of the other, one offensively and one defensively. And that is all there was to it.

There were also widespread conifer forests (pines, firs, and spruces) during most of the Mesozoic Era, along with many woodlands dominated by cycads, ginkgos, and ferns. An impressive place to be if not for all the big critters haunting the woods that liked to eat things.

Paleozoic Era

Going back further in time and further down the geologic scale, before the Mesozoic was the Paleozoic Era. Note from the Table that the Paleozoic Era began some 570 million years ago with the dawn of the Cambrian Period (toward the bottom of the table) and lasted until another, and indeed the greatest known, mass extinction of all time, the end of the Permian Period some 245 million years ago. The extinction at the end of the Cretaceous with the disappearance of the dinosaurs and many other animals and plants paled in comparison to the extinction at the end of the Permian. There is some wiggle room here on just how many species or animal families became extinct at the end of the Permian (it is impossible to really know) but most estimates are around 95 percent of all plant and animal species died out.

Whatever the exact number of species or individual animals that went extinct, life on earth came very close to ending at the close of the Permian Period. Like the extinction at the end of the Cretaceous, the Permian extinction left few clues as to its cause. Again, some scientists like the comet idea simply because it seems like a neat way to get rid of most everything in one big blast. But because the Permian Period was so long ago, any telltale signs of a Permian age crater leftover from a comet impact has more than likely eroded away eons ago or was covered up by a mile of sediment, and without hard evidence it is difficult to prove one way or another. Such as it is, there is some evidence in the fossil record showing a gradual species die-off over thousands, if not several millions, of years preceding the end of the Permian. What this suggests is neither the extinction at the end of the Permian nor the Cretaceous looks like it occurred in one day, or one year, but took thousands of years to occur, which would suggest rapid (in geological terms) climatic and sea-level changes. As the end of the Permian approached, there were worldwide continental disturbances taking place; earthquakes, volcanoes, changing weather patterns, and sea level changes, all could have, and most likely did, have a dire effect on the many land and sea-life forms.

The Paleozoic Era is made up of seven extremely long periods, the oldest, the Cambrian and the youngest, the Permian. While discussing the Paleozoic Era I'm going to begin at the bottom of the chart (Cambrian Period) and work back up Table 1 to the Permian extinction and say a few words about each.

The Cambrian Period sort of started it all, at least in terms of the fossil record. At the beginning of the Cambrian Period there suddenly appeared evidence of the first hard-shelled animals being fossilized. I can not overstate this, but a real proliferation of hard-shelled sea creatures suddenly appear as if out of nowhere in the fossil record at the beginning of the Cambrian. Prior to this there is little evidence of hard-shelled animals, and what fossils that have been found are soft-bodied animals that have been preserved in very fine shale or flint deposits sparsely scattered in various continents of the world. Animals were not necessarily rare in the Pre-Cambrian but sea creatures had not yet developed hard parts, like shells, or other hard body parts, that made being preserved as a fossil much easier.

There is some evidence of primitive sea plants in the Pre-Cambrian, and bacteria called blue-green algae. Blue-green algae lived in scattered colonies in the shallow waters of tidal pools taking up carbon dioxide from the water and giving off oxygen as a useless by-product much like plants do today. The colonies secreted a viscous substance which trapped fine sediment suspended in the waves along the sea shore, and over time the sediment would completely cover the colony, thereby necessitating the regrouping of the colony to the top of the trapped sediment. This cycle repeated itself time and again, ultimately creating large round mounds several feet in diameter and a foot or so high (the fossil sea mounds are called stromatolites). Stromatolites are among the oldest known fossils and are found in rocks throughout geologic history back to 3.5 billion years ago, and can still be found today, alive and well, in Sharks Bay, Australia.

Living in vast and scattered colonies along the shallow shore lines, stromatolites gave up trillions of tiny bubbles of oxygen into the toxic Cambrian (and Pre-Cambrian) atmosphere. For untold millions of years these ancient bacteria colonies supplied oxygen to an otherwise oxygen-free and lifeless atmosphere. Nature supplied oxygen to the atmosphere the hard way, a trillion little bubbles at a time; even so, it would be a long time in the future before the first land animal would need it. The significance of these ancient oxygen factories goes without saying, at least in terms of the future of animal life on land.

It has always been a mystery why the sudden explosion of animal life possessing hard parts occurred at the beginning of the Cambrian. Possibly the eons between the Pre-Cambrian and the Cambrian are represented by a vast time gap of little to no sediment accumulation necessary for fossilization. Possibly many fossils were preserved but during times of earth upheaval they were eroded away. Whatever the cause, by the time the Cambrian rolled around there were a multitude of relatively small sea creatures like the trilobites (Figure 4), mollusk-like brachiopods (Figure 5), and primitive sponges.

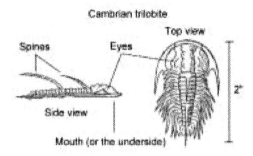

Figure 4 — A trilobite from the Cambrian Period, this particular one had spines projecting from its back and from each side of its "head." Not shown in the picture are the multiple legs the trilobite had on its underside.

The Cambrian seas were home to thousands of animal species (at least 500 species of brachiopods alone) which had suddenly appeared as if out of nowhere in the fossil record. And these animals were not slimy, little creatures floating around in some primordial soup, they were complex animals with appendages (legs, claws, antennae), mouths, eyes, and beautifully-coiled shells. All were perfectly adapted for their time and environment. Trilobites were among the first known animals that had functional eyes, in this case they were compound eyes, much like the compound eye of the fly today. It is difficult to imagine that 570 million years ago eyes had already evolved and could see quite well the food sources available on the shallow sea bottoms. What a tremendous advantage eyes would have been as a defensive weapon, enabling the trilobite to see blind bottom predators before they got too close. Maybe that's why there were so many of them.

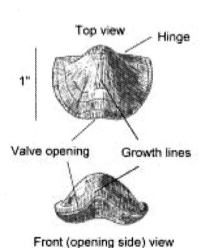

Figure 5 — Although the brachiopods looked similar to clams, they are not closely related.

Throughout geologic history the fossil record delineates the slow evolution of animal and plant species, like the fox-sized, four-toed horse of the early Tertiary called *Eohippus* to the one-toed giants we love to be near and ride today. We see it in the evolution of the dinosaurs, first seen as small, chicken-sized, half-mammal-like reptiles eating bugs and fish, only to culminate 255 million years later as the largest and meanest animals ever to live. The flip side to this manner of slow evolutionary change observed throughout geologic history was at the beginning of the Cambrian. Prior to this, in the Pre-Cambrian, are found fossils of mostly soft-bodied animals like jellyfish and worm-like creatures. In the next higher layer — now called the Cambrian Period — there is found (as I stated before) a profusion of highly evolved and complex hard-shelled animals, some even with EYES. There was no "lead in." For whatever reason, a swarm of animal groups perfectly adapted for their environment suddenly and without warning blasted on to the stage without an introduction. This early Cambrian Period event is what paleontologists (geologists who study fossils) call the Cambrian fossil explosion — and indeed it was.

During the Cambrian Period there were no land plants…anywhere. Picture a desolate earth without the first sign of a land plant as far as the eye could see, or for that matter, a land animal (all animals lived in the sea at that time). It was not that the land plants had died out for some reason, it was because no plants had ever lived on land before — ever — since the creation of the earth over 3.5 billion years before. There was not a single blade of grass on any continent in the world. Not a sign of a tree, no cattails, no ferns, no flowers, not even moss — nothing but rocks. Had it not been for the abundant water that had gathered in natural impoundments and was flowing through the rivers, I suspect the landscape of the earth would have looked as desolate and bleak as the moon. Plants would not make it out of the water for another 150 million years.

As ridiculous as it may seem, think of something right now that will not happen for another 150 million years. An eternity for sure, but not so long in terms of the age of the earth. Nature just doesn't seem to be in a hurry; could it be that Nature just enjoys the random events of every day no matter what happens? Humans are programed to remember their own past, participate in the present, and ponder their future. This is just the way we are. But we are a product *and* part of our environment (Nature) at the same time, just as the giant redwood tree or the smallest microbe. It is our nature to be and do what we do simply because hidden in us somewhere are the instructions of our behavior. We can not get rid of it or out of it, we are not outside looking in; we are born to, and inside Nature. We have evolved to breathe it, eat it, feel it, and perhaps be swallowed by it. Nature even gave us something it gave no other animal (or we gave it to ourselves), the ability to prejudge our own activities; we can plan an action before we do it. Some people use this ability to plan good and some people plan bad depending upon which cultural environment they find themselves. What may have been a good choice for a Neanderthal 150 thousand years ago may be a bad choice today. But it does not make the slightest bit of difference in terms of our natural behavior simply because by nature we are allowed to chose and then live with the consequences.

Humans are a product of evolution (like everything else), nothing we do can be construed as unnatural or unacceptable except in one's own specific cultural setting. Roman culture established as natural behavior certain liberties we might find abhorrent today, so did the Mayans, and the Neanderthals — it worked for them. Actually, we can not do anything Nature does not want us to do. Humans can not spread their arms and fly, they can not breathe water, nor can they run as fast as a cheetah — these feats are unnatural for us but they seem easy for birds, fish, and cheetahs. And although all animals (it goes without saying) are indeed conscious, humans are conscious that they are conscious; it seems this extra little flair sets us apart from all the other animals on earth. And with this gift we can think and then do what comes natural (whatever that may be). Bad choices during our several million years of evolution usually meant death which, of course, was certainly all right with Nature but taught the surviving Klan not to make that same decision again. Contrary to this, making bad choices over the last hundred years or so just got us in trouble for a while but we could usually go on and make more bad choices if we wanted to and get in more trouble. Our culture has become more obliging to bad choices, at least as far as death is concerned. Care must be exercised though, because there is a great difference between being *au natural* and being "normal." In today's society, like all past societies, cultural decree rules.

But, excluding humans, animals don't have culture; yet, they do have social order and instinct. And instinct is nothing more than a natural reaction to some external stimulus that has evolved, not through trial and error, but through natural differences among the same species. An example would be during the evolution of rabbits some liked to sit out in the open and watch the hawks fly by, some didn't. After so long a time, only the rabbits that didn't like to watch hawks were left to have more rabbits that didn't like hawks just like their parents. This wasn't a learned trait but just a matter of an inherent difference of opinion.

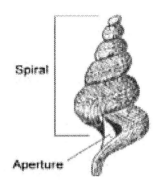

Figure 6 — An Ordovician gastropod (snail) with a high spiraled shell. The snail carried the shell horizontally on its back with its body snug in the aperture.

The Cambrian seas held a multitude of primitive plants and sea-dwelling animals. Because of the abundance and varieties of trilobites, the Cambrian Period is often referred to as "the age of trilobites" (remember, they could "see"). The Cambrian lasted a long time and many animals and plants came and went, but at its close and into the Ordovician Period, there was an even greater expansion of life in the sea. More than twice the number of species have been identified in the Ordovician-age rocks than in the older Cambrian.

Many species made great advances during the Ordovician Period (not only in the number of species but in complexity), like corals, trilobites, gastropods (Figure 6); cephalopods (squid, octopus, etc.) some of which are the coiled-shelled ancestors of the modern day *Nautilus* (Figure 7); pelecypods (bivalves, or clams) and brachiopods.

When in the field collecting fossils the most obvious way to distinguish the difference between a gastropod and a cephalopod is the gastropod spirals upward around a central axis and the cephalopod coils around a central point. During the Ordovician, the brachiopods finally outnumbered the trilobites, and during the middle and late Ordovician, peculiar, boney, and jaw-less fish began to appear in the fossil record. Plants were still an inhabitant of the sea; true algae, primitive sea weed, and other early plant life must have been flourishing during this period but plant fossils representing the

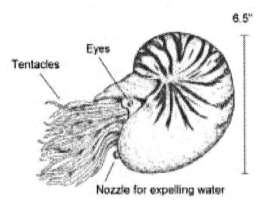

Figure 7 —The living *Nautilus* is the last surviving member of a very large family of coiled-shelled cephalopods that first appeared in the middle of the Cambrian and all but went extinct at the end of the Cretaceous Period.

Ordovician are few in number. Because of the number and diversity of the brachiopods, the Ordovician Period is sometimes referred to as the "the age of the brachiopods," even though there were plenty of other animals in the oceans as well.

The next younger period, the Silurian, saw a decline in trilobites. Corals continued to develop into many varieties, both solitary and colonial. Brachiopods stayed about the same, however, evolving into more elaborate and diversified forms. Fish experimented with various evolutionary designs during most of the Silurian, but toward the latter half of the Silurian Period they began an explosive evolutionary progression of many different types and numbers that continued throughout even the next period.

The first known air breathers (or land dwellers) finally appear in the fossil record in late Silurian-age rocks. These air breathers were represented by flightless insects and scorpion-like arachnids (ancient kinsmen to today's spiders). The insects didn't have lungs (and still don't today) so they didn't really breathe. They had tiny holes in their sides that allowed oxygen to enter their bodies and transfer to their circulatory system — but they still count as air breathers. Imagine that, the first living creature to set foot on the earth's dry land was a bug. There is some irony in this; even if you lived some 420 million years ago you would have had to cope with primitive, crawling insects and ancient spider-like creatures. These primitive animals were, of course, confined to the water's edge since there was little on land for them to eat or bite, unless they ate each other.

Plants also finally made their way to the land during the Silurian Period, although with very simple and rudimentary vascular designs (circulatory systems, roots, stems). For the first (first means during the Silurian) time ever on earth a little green began to slowly appear around the rocky coastline and calm tidal pools (excluding the stromatolites that were sometimes exposed during low tide). These ancient plants were more like small, pale green, soft-tissued, leafless "sticks" poking up above the sand and between the soaked shoreline rocks. This first step was a big one though, because it allowed plants to ever so slowly begin to adapt away from a total water environment; first by developing stronger stalks to take the body of the plant off the rocks, second by adapting their root system so they did not have to rely on total sea water but could probe down into the ground for the moisture they needed, and third by developing stalk projections (primitive leaves) to increase their surface area thereby increasing food processing areas.

In today's world, as it was in the Silurian Period, there are basically two types of animal skeletal designs, the invertebrates and the vertebrates. The first, the invertebrates, are all the animals *without* backbones, such as the molluscs (clams, octopuses, etc.), insects (flies, beatles, mosquitoes, cockroaches, etc.), spiders, worms, snails, coral, jellyfish, lobster, and so on. The second type, the vertebrates, are animals *with* backbones. Like the invertebrates, there are thousands of vertebrate species, such as horses, elephants, birds, dogs, cats, fish, whales, rats, bats, people, snakes and so forth. The differences between the haves and have nots were becoming more apparent during the latter part of the Silurian Period and into the Devonian Period.

Most invertebrates have some type of external skeleton to hold their organs together, like shells (calcium carbonate) or chiton (fingernail-like material) which provide a type of body armor and appendage covering. Having an external skeleton served a couple of purposes; it provided a way to keep the body from just being a glob of body gumbo, and it provided some protection from other animals that liked to eat gumbo. With the exception of fish, most seafood we consume today is the invertebrate type. No one has ever bit into a bone while eating lobster because lobsters have no internal skeleton (bones). Once you break through the lobster's external chitinous skeleton the rest is meat, or so they say. The same is true with clams, oysters, and crayfish. Get rid of the shell and the rest is boneless.

Having an internal skeleton, on the other hand, certainly provided support for the body and something for the muscles to attach to so the animal would not fall apart; however, it did little for their external protection. So vertebrates had to devise other means of protection, like being able to flex the backbone and swim faster than the other animals trying to eat them, or developing teeth so they could bite back if need be. One other important advantage of the internal skeleton was that it allowed uninterrupted growth. Many animals with external hardware could only grow so big and then they would have to "molt" or discard their hard exterior covering and grow another one, like today's crayfish and yesterday's trilobites. During these periods the animal was more vulnerable to predators. This is not true with clams, oysters, gastropods, and such, whose external shells just continue to grow. Whatever, as I mentioned earlier, the vertebrates, mainly in the form of primitive fish, became more numerous and diverse in the late Silurian seas.

Fish were not the first animals with backbones (although primitive fish get most of the credit). Many vertebrate creatures obviously preceded them. Small (about two inches long), flat, worm-like creatures (called chordates) found in the middle Cambrian show visible signs of rudimentary backbone-like structures 90 million years before ancient fish sprang onto the scene, and there are a lot of unknowns in between. I really do hate to refer to these small chordates as worm-like because they were not worms at all, indeed they are the earliest fossils found to date of animals with a hint of a backbone. Most likely fish as we know them in the Silurian seas evolved from these primeval, and very special creatures.

After the Silurian came the Devonian Period (see Table 1). The vertebrates gained dominance over all other aquatic creatures — not so much in terms of numbers but in terms of size and their predatory nature. Although the fish were still primitive, they had evolved into several more specialized types. There were still the jawless types getting their nourishment by sucking up organic sediments from the sea floor or straining small floating plankton from the water itself, there were "platy-skinned" fish that had developed workable jaws, some of which reached a size of twenty feet or more in length (a fearsome looking animal for sure), there were primitive ancestors to the modern shark, and there were what we might think of as true bony skeleton fish. And for the first time ever, during the late Devonian, vertebrates finally crawled upon land.

I think in this case "crawled" is a good word because the first known animals (beside insects) to make the transition from water to land were the amphibians and lobe-finned fishes. The lobe-finned fishes had very primitive lungs and could breathe air for a while when the tidal pool in which they were living began to dry up. It is suspected they could "walk" from one pool to another while looking for food. Most likely the lobe-finned fishes were also adapted to freshwater and on occasion it was advantageous to leave and seek out new water pools.

Lumbering flat-headed amphibians — who probably had the same common ancestor as the lobe-finned fishes — must have pulled their wet and slimy bodies from the water regularly and gobbled up some of the abundant insects and spiders, then went back in the water to lay their eggs. In contrast to the lobed-finned fishes which had four strong boney fins to support them on land, the amphibians had four primitive legs and adequate lungs to breathe as much Devonian air as they wanted. Had it not been for the stromatolites supplying oxygen to the atmosphere for the previous two billion years or so, there would have been little need for the lobe-finned fishes and amphibians to develop lungs, primitive or not.

Brachiopods reached their zenith during the Devonian and attained great diversity in form. Corals and molluscs (gastropods, cephalopods, pelecypods, etc.) also enjoyed great success. Trilobites, however, were in decline in spite of the fact of gaining great size and diversity in some species (up to two feet long).

Plants during the Devonian, especially by the middle Devonian, began to distance themselves a little from the sea shore and proceeded to move slowly inland. Small, leafless, "scale-like" plants and primitive ferns (with leaf-like structures), both true ferns and seed ferns, produced heavy (if not low) growth on what was once barren rock — 385 million years ago the earth was slowly turning green. Grasses had not yet made their appearance and wouldn't for at least another 320 million years during the early Tertiary Period. What a sight the late Devonian must have been; amphibians and other lungfish struggling in and out of water pools and rocky-bottom, shallow rivers, some trying desperately to navigate through the dense, low-lying greenery along the waterfront, all the while stirring up and snapping at insects and whatever else was scurrying about. The Devonian seas were flush with shellfish, fish of all sorts, jellyfish, coral, sharks, trilobites, and cephalopods. It was also the time of the first true insects (wingless) and more varieties of spiders, scorpion, and millepedes.

During each of the Periods so far discussed, not only were the animals slowly changing but the continents were also changing, not by day, not by year, but by the millennia. Fluctuations in sea level, mountains imperceptibly being eroded to broad, grass-less prairies, flat plains and swamps being raised to mountains, shifting weather patterns, and so on caused many animal groups too specialized to adapt to die out, and others not so specialized to succeed.

The animal groups that survived were not necessarily "more fit" in this case than the animal groups or species that died out. All animal groups were perfectly adapted to the time, place, and climate in which they lived; if they hadn't been they wouldn't have been there in the first place. It was just as easy for a more "advanced" group of animals to come upon hard times as it was for what we might call a less advanced group. When an animal group evolves during long periods (millions of years) of plentiful, say, bamboo trees, and this animal's primary diet happens to be bamboo leaves, if the climate changes rather rapidly and the bamboo trees die out so will the animal (not unlike what is happening to the Panda bear today because of urban sprawl). You just can not get those stubborn Pandas to eat anything else.

Specialization is good when your favorite food is handy and plentiful but not so good when that particular food source begins to dwindle. Especially if the food source dwindles rapidly over several thousand years or so. This may seem like a long time but it's not long enough to adjust to eating other foods when your physical make up and predisposition relies on that one food source. On the other hand, if an animal is scurrying about in the undergrowth eating anything it can get its claws on, which might include leaves, nuts, grass, worms, insects, and anything else, the chance for survival during a rapid environmental change is greater for this animal. When one food source becomes scarce, it can always scrape up a meal somewhere else. Sort of like not putting all your eggs in one basket. That was what was happening during the entire Paleozoic Era, time and time again some animals did well and some did not. Could it be said that the animals that were doing well were just lucky?

The Period after the Devonian, and next younger, was the Mississippian Period. The Mississippian lasted roughly 45 million years and saw many different climate changes. For long intervals, West Virginia was only a few tens of feet above sea level as part of a low, sinking land trough that extended from Alabama northeastward to Newfoundland. Great quantities of sediment were being eroded from the old Appalachian (Acadian) Mountains to the east (forerunner to today's Appalachian Mountains). Huge quantities of sand, silt, and mud washed down from the mountains and were deposited in the broad, flat trough (called a geosyncline) by sluggish, meandering rivers. These sediments formed deep deposits of clastic sediments. "Clastic" refers to rock fragments and particles weathered and broken off of preexisting rocks and then deposited somewhere else. The preexisting rocks in this case were the rocks of the Acadian Mountains.

The rock fragments, washing down from the eastern mountains, came in all sizes depending upon the amount of water washing them down at any one time; from pebbles on the large side to clay sized particles on the small side. Thousands of feet of sediment poured into this sinking, low geosyncline. Just one series of red shales (mud) alone deposited in West Virginia during the Mississippian Period was several thousand feet thick.

The Mississippian Period was also a great limestone forming period. Several times during this period the sea inundated much of West Virginia forming thick accumulations of marine (another name for the sea or ocean) shales and limestones. The sea advanced through sea lanes (flat corridors) which were opened from the Gulf of Mexico up through Texas, Missouri, Tennessee and into West Virginia, from the north it came down from western Canada, through Idaho, Nebraska, all the way into West Virginia. There was a lot of sea water covering North America from beyond Canada to the Gulf of Mexico and almost everything in between.

The "continental" sea was, for the most part, warm and relatively shallow and teeming with all sorts of marine life like foraminifera, tiny planktonic, calcite-shelled sea creatures; bryozoans, small coral-like animals that grew in colonies from a branching or fan-shaped structure (Figure 8); shellfish, brachiopods, fish, and coral just to name a few. Most of the North American interior was so close to sea level during the Mississippian that a rise in sea level of just a few feet (or the subsidence of the land of a few feet) would inundate thousands of square miles of the continent.

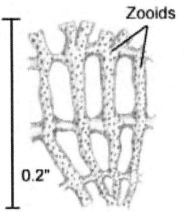

Figure 8 — A section of a bryozoan colony. Each heavy dot represents an individual living chamber, called a zooid, of the tiny bryozoans.

Limestone was formed when the sea covering West Virginia was at sufficient depth for the accumulation of lime ooze deposited by marine organisms, or was accumulated from direct "lime" precipitation from the sea water itself (Chapter Two). Several thick deposits of Mississippian limestone, including the Greenbrier Limestone, lie about a thousand feet below the Kanawha Valley. West Virginia's official state gem is not really a gem at all but instead a silicified Mississippian-age fossil coral found in the Hillsdale Limestone in Pocahontas and Greenbrier Counties. Going by the rather lengthy name of *Lithostrotionella*, when polished this ancient coral colony is indeed gem-like and makes handsome jewelry and (to some) an impressive coffee table display. The Mississippian limestones also hold most of West Virginia's famous caves, and supply much of the crushed rock used in road building in the eastern half of the state. All the limestone was deposited when the sea covered much of the West Virginia's sinking geosyncline for countless years, depositing thick layers of limey mud. In Chapter Two an explanation of how limestone and other rock types form will be provided.

Plant and animal life during the Mississippian progressed in a somewhat unspectacular but ambitious rate. Land plants acquired well-developed root systems, improved leaves to convert sunlight into food, and more efficient circulatory (vascular) systems. These developments allowed the plants (I guess I can start referring to some of these plants as trees) to grow bigger and higher, lifting themselves higher off the ground to take advantage of the space above — the more space, the more leaves that could be produced and exposed to the sun. Soft tissue scale trees (so called because of the scalelike impressions left on the tree's outer surface by detached leaves) grew to bewildering heights — some over one hundred feet — giant relatives of the modern horsetails (scouring rushes) grew 30 to 40 feet tall, primitive seed ferns (now extinct) and true ferns (ferns that reproduce with spores like modern ferns) grew up to 50 feet tall, and a multitude of ground-hugging, herbaceous (bushy) plants flourished. Could it be that sometime during the early Mississippian Period a small primeval tree cast its cool shadow over the hot, rocky landscape for the very first time?

Although this sounds good, the answer is no. There never was a time when one day there was no shadow and the next day there was. Each day, each year, each thousand years, plants were plants and animals were animals. The incredible slowness of evolutionary change can be seen only in the extreme — from one time to another. What we see today as fossils are a very small part of the plant and animal lineages that actually lived during these periods. The fossil record shows us only bits and pieces, with so much missing in between. So, it is easy to say that the first insect ever found that lived on land was during the Silurian Period (which is true), but who knows how many millions of years the insects slowly perfected this feat and who knows how they did it. A Silurian insect with air tubes in its sides could thank Mom and Dad because they had air tubes as well.

The often-asked question, "Which came first, the chicken or the egg?" can doubtless be answered, "neither" because there never was a time when one day there were no eggs (or chickens for that matter) and the next day there were. There never was a time when one day there were no birds and the next day the red-wing blackbird appeared and discovered how much fun it was to perch sideways on the stalk of a cattail. As mentioned earlier, every living plant and animal we see today has an ancestry going back to the beginning of life. Somewhere eons ago (possibly during the Mississippian Period) the ancient ancestors to the birds (which includes chickens) laid clusters of soft-skinned eggs in the water much like amphibians do today. Even before the Mississippian, these creatures had ancient ancestors all the way back through the Cambrian and into the Pre-Cambrian where naked embryos were just cleansed from the body to fall or drift where they may; some made it and most did not. And even these creatures had ancestors where reproduction was just a matter of a hunk detaching itself from the "parent" or the "parent" just splitting in two! And, you guessed it, even these creatures had ancestors. So, which came first; follow the slow evolution of any organism back through geological time and the word "first" has little meaning.

Think of the bewildering odds that each one of us is here at all; had just one of my ancestors, all the way back to the Pre-Cambrian seas, lost a fight, got eaten as a baby, or just decided to remain single, I would not be here. Each of us is connected to the past simply because our past has never been broken. We are the product of millions if not billions of ancestors, which as far as we are concerned made the right choice. Those plants and animals we do not see today also had an unbroken genealogy from the beginning of life until their time ran out.

Primitive trees were indeed bigger during the Mississippian Period and casting their shadows all over the place, but how many eons it took the shadow to go from nothing to something nobody knows. Adaptation does not come easy; there may not have been any real "firsts" in the prehistoric past but there were plenty of "lasts."

During the Mississippian Period, one treelike plant was slowly developing a stem (trunk?) a little different from all the other relatively soft-tissued trees. This primitive plant was developing real woody tissue. The ancient ancestor of the conifer (evergreen) was slowly evolving. Although it was not much during the Mississippian Period compared to today's lofty pines, hemlocks, and towering redwoods, it was, in a word, the forefather of the true "woody" trees we look up to today (those still standing).

Mississippian animal life also flourished. On several occasions broad, shallow seas covered much of the continent which allowed rapid development of the vertebrates. Fish were freely evolving into many different species and doing quite well even in freshwater. Primitive sharks with special platelike teeth used to crush shellfish were having a heyday. Trilobites and brachiopods were still numerous but in decline, and clams and snails were common in marine and freshwater environments. On land, of course, the insects and spiders were multiplying as always and getting bigger. The first winged insect was found in Mississippian-age rocks. Amphibians and similar land lubbers became more numerous and more adapted to a life on land. It took a long time but the amphibians and lungfish were finally doing quite well out of the water. Many traces (tracks left in sandstone) of amphibians have been found in Mississippian-age rocks in West Virginia. It seems strange to say, but during the Missippian Period the amphibians, lobed-finned fish, insects, and arthropods (spiders and their relatives) were the dominate land animals. The whole world was theirs for the taking.

After 35 million years the Mississippian Period at long last came to an end. The boundary between the end of the Mississippian age and the beginning of the Pennsylvanian Period is one of erosion. Regional uplift of the heretofore sinking trough exposed some of the top, or youngest, beds of the Mississippian to the elements, wrenching from their burial site the freshly laid and exposed sediment once again. After a lost time span of erosion, the trough was again lowered to an area of deposition which then accepted the new deposits of the Pennsylvanian Period.

And, finally, we come to the Pennsylvanian Period, named because the first work done describing the rocks of this age was done in, of course, Pennsylvania. Those who have diligently followed Table 1 and looked ahead when I first said the rocks of the Kanawha Valley were around 300 million years old know that they must be of Pennsylvanian age. And that is true but the Pennsylvanian Period lasted some 40 million years, and a lot can happen in 40 million years. Just how long is 40 million years? I have no idea, but an incredible long time indeed. We do know it was long enough to deposit almost a mile of sediment consisting of many massive sandstone layers, alternating shales, clays, multiple beds of coal, and of course, fossils. The 40 million year duration of the Pennsylvanian Period occurred one day at a time, nothing noticed. Each day was a new contest, some animals and plants made it through the day and some did not. Are the rocks exposed around Kanawha Valley in here somewhere? Yes.

The rocks of the Kanawha Valley "fit" into, or were formed during the Pennsylvanian Period, and since the rest of the book will be about this particular period a word or two about the following Permian Period so the discussion about all the periods will be complete.

Sediment deposition all but stopped in much of the Appalachian Geosyncline in the early Permian with the exception of a rather small (about 8,000 square miles), isolated area in what is now mainly the Ohio Valley covering parts of western West Virginia, eastern Ohio, and western Pennsylvania. These Permian beds are called the Dunkard Series but have no outcrop in the Kanawha Valley but do outcrop as red sandstones and shales along I-77 through Jackson and Wood Counties.

Although life was flourishing in the Permian, it was also a time of stress. In Chapter Three we will discuss some of the continental changes that were taking place during the Permian Period which caused the stress. There were many climate changes during the Permian and although cone-bearing conifers were doing well because of the changing climate and increased aridity, many plants were actually dying out or becoming smaller. During the Permian the *Ginkgo* (maiden-hair) tree first appears in the fossil record. The *Ginkgo* has remained for the most part unchanged since the Permian Period and represents the oldest, unchanged tree form in the world today, a true living fossil. Look around the Kanawha Valley, there are many *Ginkgo* trees planted as ornamentals and shade trees. They can reach heights of 50 or 60 feet, a magnificent tree that has perfect, fan-shaped leaves.

Trilobites were just about gone and would be by the end of the Permian. Brachiopods, cephalopods (most would be extinct at the end of the Permian), pelecypods (shellfish), gastropods (snails), and corals were doing good; insects were getting smaller and four-legged reptiles were getting bigger and more varied. Most reptiles of the time had evolved stocky bodies and four short legs, some had large fins on their backs (*Dimetrodon*). Half-mammal, half-reptile creatures plodded about; true mammals had yet to make an appearance — at least in the fossil record. But, as discussed above, there was an impending doom coming at the end of the Permian; the animals and plants did not know it but times were slowly becoming more difficult, temperatures were rising and rain fall becoming less. Many changes were going to occur between the latter part of the Permian and the next period, the Triassic, and most species would not live to see it. The greatest of all known extinctions was looming just over the horizon.

So far we know some of the events that occurred after the Pennsylvanian Period, such as the Permian extinction and the dinosaurs of the Triassic, Jurassic and Cretaceous Periods; the large mammals of the Cenozoic Era and ice sheets, and we now know some of what occurred before the Pennsylvanian Period, such as the Cambrian fossil explosion, the Ordovician fish, and the first flightless insects of the Silurian. We covered 570 million years in just a few pages but I hope these pages gave you just a glimpse of some of the changes that have occurred to our physical planet and to the life it harbors.

Because the rocks around the Kanawha Valley are of Pennsylvanian age and because they were already here when the dinosaurs stomped over them, you will not find any dinosaur fossils in the Kanawha Valley as hard as you may look. The rocks we see exposed in the Kanawha Valley today were already here for at least 100 million years before the great epoch of the dinosaurs. As a matter of fact, you will not find anything, animal or plant, that lived before or after the Pennsylvanian Period in the rocks of the Kanawha Valley, but you will find the animals and plants that lived during this period when the rocks were being deposited, and what a grand collection it is. The Pennsylvanian age was a great time for both animals and plants, and their story is well represented in the rocks of the Kanawha Valley.

To recap, we have covered the three described Eras and the 12 periods of the geologic time scale, roughly 570 million years of earth history. Although this is a long time even in geologic terms, it is only about an eighth of the actual age of the earth. There are no words to even begin to describe the time past in the Pre-Cambrian; it is an abstraction so I will not try. But somewhere back there life began. The earliest life form known was simple bacteria, and, of course, there was obviously something before that. Whatever started it all will probably forever remain a mystery simply because it was so long ago that most of these ancient rocks have eroded away and what life was there was not easily fossilized. If they were fossilized, their impressions have become obliterated just because of the bewildering age of the rocks. There was no birthright, no expectations, no remorse, only chemical familiarity.

So, the rocks of the Kanawha Valley are of Pennsylvanian age, and now when the reader rides through the valley and sees the large exposures of rocks along the road he or she will know where they fit into the overall geologic record, much happened before and much happened after these rocks were laid down. In the next couple of chapters I will "zero in" on the Pennsylvanian Period and provide more detail on the rocks of this period and the fossils found in them. Remember, only animals and plants that lived in a particular period can be found as fossils in rocks of that period. When a fossil is found in the rocks of the Kanawha Valley it represents life as it was in the Pennsylvanian Period some 300 million years ago.

Before I start the next chapter I want to explain, as best I can, something that may be a little difficult to grasp. I am saying things like, "At the beginning of the Cambrian Period there was a fossil explosion" or "There was the greatest extinction of all times at the end of the Permian Period." This makes it sound like all of the periods were named already and written in stone by some ancient prophet and these occurrences just happened to take place coincidentally with the beginning or ending of certain periods. That is, of course, not the way it happened. The fossil explosion was obviously already there, buried in the rocks, long before anybody discovered the fossils and named the periods. What scientists have done is identify certain significant events in the geologic record, like, again, the fossil explosion, certain extinctions, certain land or mountain-forming events, and so on. From these events they (the scientists) drew the lines of the periods as well as the eras, the eras being more significant events than the periods. As an example, the great extinction at the end of the Permian Period marks the boundary between the Paleozoic Era and the Mesozoic Era. So, in a word, the story told by the rocks determined where the lines were drawn, not the other way around.

Most of the period names represent an area where the rocks of that age were first studied. For example, the Jurassic rocks were named after the Jura Mountains located between France and Switzerland where they were first studied; the Permian was named after the province of Perm in Russia; the Cambrian, named after Cambria, the Roman name for Wales where it was first studied; and as already mentioned, the Pennsylvanian named after Pennsylvania.

In the geologic scale (Table 1), most of the European countries combine the Mississippian and Pennsylvanian Periods into one big period called the Carboniferous Period, so called because of the coal (carbon) deposits of these two periods throughout Europe. With this in mind, on page 27 I mentioned the K-T extinction boundary (clay layer) that marks the passage from the Cretaceous to the Tertiary Period and that a "K" was used instead of a "C" to avoid confusion with another period, the Carboniferous Period is the other period.

Chapter Two
Stone Story

For the frivolity of my mind Lord,
don't blame me.
It's natures way,
can't you see.

The Pennsylvanian Period lasted some 40 million years. At the dawn of this period the earth was already green with thousands of different plant species ranging from ground-hugging herbaceous (bush and vine plants) to towering trees, and just as many insects, spiders, cockroaches, and scorpions. Nature must have loved beetles above all else because during the Pennsylvanian Period, as today, there are more beetles and beetle species than most other species combined. Primitive sharks swam in the sea feeding upon the many varieties of ancestral fish. Amphibians reached great size and variety, with one species ten feet long and weighing up to 500 pounds. Picture that crawling out of the farm pond. The amphibians were the king of the land beasts during this period as the sharks were in the seas. And by the end of the Pennsylvanian Period would be the greatest accumulation of plant matter the world has ever known, and the very first small, reptile-like creatures would make their appearance. What a marvelous 40 million years it must have been.

It seems the Pennsylvanian Period was a time of extravagance for both plants and animals, everything got bigger and more prolific. It was also a time of great accumulations of sediment. The broad, sinking trough that covered much of West Virginia during the Mississippian Period continued throughout much of the Pennsylvanian Period to be the recipient of enormous quantities of the sand, silt, mud, and clay that was still being eroded from the ancient Acadian Mountains to the east. And, for the most part, West Virginia, especially the Kanawha Valley, was as flat as a board.

As we look around and up to the hills it seems wholly unlikely that the Kanawha Valley was once part of a boundless lowland crossed by hundreds of sluggish, meandering, sediment-choked rivers, or at other times part of a vast swamp that stretched into Ohio, Kentucky, Virginia, and Pennsylvania. The Kanawha River was not here, neither was the Elk, Gauley, or New. The rivers that flowed through the "valley" had no names, no lore, no songs written about them. They were just rivers. But they were not rivers like we think of today, flowing along in one nice, continuous path, like the Kanawha River, ever deepening its channel and at the same time trying to erode its banks while it swings back and forth across the flood plain. The rivers meandering across West Virginia during the Pennsylvanian Period were not eroding anything; they were in fact depositing stuff that had been eroded many miles to the east. With each rain, sediment-choked water poured off the eastern highlands in a thousand different streams, spilling together in the flats forming broad, aimless, and braided rivers of sediment. Not just one but hundreds of newly-formed rivers squirted from the entire length of the western boundary of the foothills, the coarse sediments coming to rest where the highlands met the lowlands while the finer material continued into the open and shallow expanse or was carried by the meandering rivers into quiet, backwater pools. At times the sand deposits formed huge deltas that extended many miles into the trough, and as the trough slowly subsided the sand deposits just got thicker.

There are many instances in the geologic past where these vast sinking troughs, or geosynclines, existed. As the land ever so slowly subsided, they were maintained at about the same elevation by more and more sediment filling in the low spots, and the low spots were hundreds of miles across. Over millions of years thick deposits of sediment can accumulate in a geosyncline. The total deposits of the Pennsylvanian Period alone are almost four thousand feet thick. That was the story during much of the Pennsylvania Period, but also there were times when the broad flat Appalachian Geosyncline became nothing more than a vast, shallow swamp stretching for hundreds of miles in all directions. Dense swamp-forests, marshes, and peat bogs like the world has never seen before or since completely covered the states mentioned above like one enormous, soggy, insect-invested green carpet. Giant dragonflies with a wing span of up to two feet buzzed (roared) in and out of the decaying underbrush looking for a meal. Cockroaches up to four inches long, in company with five-foot long millepedes, scurried about the thick, damp scrub and stinking bottom vegetation. The nights were hot and the days were hotter and it was ceaselessly humid, an absolutely perfect environment for rapid plant and animal growth.

There were no winters to speak of during the Pennsylvanian Period because West Virginia, or the area that would one day become West Virginia, was located very close to the equator. But continental drift and plate tectonics are another story. The Pennsylvanian Period would have indeed been an etymologist's *and* a botanist's utopia, not unlike the Amazon River Basin is today.

It is hardly possible to describe the difference between our spot on the planet today and what it was like during the Pennsylvanian Period, or any of the other periods for that matter. Widespread environmental or life form change can only be perceived when looking at very extreme intervals. Nevertheless, on a local level the rocks do show many occasions when rivers suddenly changed direction, or where sand had been deposited for many years suddenly clay is deposited immediately over the sand, which in turn may be covered by several feet of coal, and so forth. The word "suddenly," of course, could mean anything. What appears to take place suddenly in geologic time by just observing the apparent rapid changes in the rock types revealed in a road cut may have taken many years to occur.

I have mentioned on occasion the terms sediment and deposition, which are fairly common terms most are familiar with. Defined, sediment is "solid material that has settled down from a state of suspension in a liquid." This definition is specific to liquid (like water) but as far as we are concerned here in this book sediment is the rock fragments (particles) and other debris (organic material) that settle or drop down from any medium, be it water, wind, or ice (glaciers). Deposition, on the other hand, is the actual act of laying down (depositing) the sediment by any natural agent, as just mentioned, like water, wind, and ice. When glacial rock fragments are deposited the final accumulation of material is called "till." No till deposits in the Kanawha Valley.

When water flows fast, say in a river flood, it is very powerful and can carry sediment of all sizes, big and small. When the water begins to slow it can't carry (or push along) the larger particles any longer, these larger particles fall out of suspension or can no longer be pushed and are deposited on the channel bottom or wherever the flood waters may be. As the water slows even more, the next size particles begin to settle out and are thus deposited in turn. This continues until the water almost stops, or stops completely in a natural impoundment or quiet backwater area, finally allowing the smallest particles to settle out. This final collection of very small particles is the fine mud or clay size material. Actually, the water, when reaching the lowlands or beginning to slow down for some other reason, literally sorts the sand, silt, and mud in terms of size; the bigger (coarser) particles here, the middle size on down river, and the fine clays over there.

This process can be easily duplicated by shaking up a handful of dirty sand or garden dirt in a jar of water with a lid on it. While the water is being shaken (moving fast) all the sediment will be suspended in the water bouncing here and there. When the shaking stops and the jar is put down at rest, the big stuff, if there is any, falls to the bottom almost immediately while the smaller and smaller sediment settles out in turn ever so slowly. When the water on top finally becomes clear, notice the different layers of sediment in the jar. The coarser particles will be on the bottom, the next size smaller will be above that and so on. The finest and smallest particles will be the clay size sediment and this will be on top. You may have to wait a while for the clay sized particles to actually settle out. This is why when you look at a rock, especially a sandstone where the grains can easily be seen, all the grains or particles making up the sandstone look about the same size, although this is not always the case as will be explained later.

BOB KESSLER

Some Stone Stuff

For the moment, let's consider the different kinds of rocks found here in the Kanawha Valley, all of which are sedimentary rocks. Sedimentary rocks are rocks formed by the deposition of some type of sediment, and sediment constitutes some type of solid material that has come from someplace else, either by water, ice, or wind, and been deposited (in the case of water, settled out). The rocks of the Kanawha Valley have all been deposited by water in one manner or another, most by terrestrial (land) sources like rivers, streams, swamps, etc., with a few scattered marine deposits. Old rock fragments eroded from the ancient, eastern highlands made great sediment. Every particle in every rock in every hill in the valley is just visiting, each came from someplace else and each will ultimately be washed away to another place. Nothing can stop it.

As you drive from Marmet to St. Albans (or anywhere else in most of West Virginia, Kentucky, Ohio, and Pennsylvania) you will see five basic rock types: sandstone, siltstone, shale (clay), limestone, and coal. There are, however, many different ways these five rocks types can be found in combination with one another. A few examples might be such types as sandy shale, argillaceous (clayey) limestone, carbonaceous (dark organic) shale, calcareous (limey) sandstone, and so forth. Most people who study rocks prefer, for whatever reason, to use big words instead of small words. Argillaceous is used to describe a rock that contains a high percentage of clay, carbonaceous means something with a high percentage of organic (carbon from dead plants) material — almost always a dark rock because carbon is black — and calcareous means a rock with a high percentage of calcium carbonate. Although I used the word limey in parenthesis above, simply because limestone is made from re-crystallized calcium carbonate, a rock described as a calcareous sandstone would actually be one that had the individual grains cemented together with precipitated calcite — calcite "glue" if you will. More about this later.

The terms sand, silt, and clay are used to describe a range of particle sizes and not to describe a certain kind of mineral in the rock. Sandstone does not have to be made up of any particular type of particle (like quartz), only a particular *size* of particle. For instance, sandstone means a rock made up of particles which are less than 0.08 in. (It would take 12.5 of them laid out in a row to make an inch) down to 0.0025 in. (400 to the inch). So sandstone, depending on the particle size, can be described as a course-grained sandstone, a medium-grained sandstone, or even a fine-grained sandstone. Silt (which forms siltstone), on the other hand, has particles smaller than the 0.0025 in. down to 0.00016 in. (6400 per in.). You can see that the silt sized particles can be quite small. Since silt is so small anyway, a silt-stone is just a siltstone (no coarse or fine-grained siltstone). Clay sized particles are defined as any particle smaller than the lower limit of the silt. In fact, most clay sized particles are so minute they can only be seen with an electron microscope, and being this small they will only settle out of suspension when the water is quite still for long periods of time.

Now, nobody goes out in the field and measures this stuff to see what type of rock it is, at least nobody I know. The point I wanted to make was that these terms refer to particle size and nothing else, and I have no idea who first described the limits between them. Someone with a lot of time on his or her hands, I bet. If the rock feels good and gritty and the particles can be clearly seen it is a sandstone. If the individual particles can't readily be seen but it still feels a little gritty it is most likely a siltstone. If the particles can in no way be seen and the surface is smooth and slick feeling it is some type of compacted clay. Some of the rocks made predominantly of clay are shale, claystone, and mudstone.

Sandstone

Everyone knows what sand is — it is the white, grainy substance scattered all over the shoreline at Myrtle Beach (small, well-sorted quartz particles), it is the tan to buff-colored material that blows in the wind and forms the big dunes in the Sahara Desert (also small, well-sorted quartz particles), and it is the multi-colored stuff that forms the steep and massive outcrops all along the interstates throughout the Kanawha Valley (mainly small quartz particles). The only difference between these three examples is, in the first two mentioned, the sand is loose; in the last example, the sand has been consolidated and all the countless individual grains cemented together to form stone (Plate 1).

Plate 1 — A road cut on Piedmont Avenue in Charleston showing several massive layers of the Winifrede Sandstone separated by thin, silty shale beds. Notice the black vertical streaks caused by minerals in the groundwater running down the face of the cut.

The sand making up the sandstone layers seen in Plate 1 was also loose when first deposited by moderately flowing rivers into and across the Appalachian Geosyncline 300 million years ago. Had anyone been there at the time they could have walked in it and left footprints, built sand castles, or maybe just sat by the riverbank and watched the big ugly amphibians lumbering in and out of the water or the primitive, four-legged reptiles plodding about looking for washed-up, dead fish. Maybe no person was there to leave their tracks but tracks of some of the ancient reptiles have been found in West Virginia preserved in solid stone.

There were no birds, so birdwatching was out, no monkeys or crocodiles, and no flowers to marvel and smell since flowering plants would not evolve for at least another 100 million years in the future Cretaceous Period. There was no such commodity as fruit trees since these all possess flowers; no maple trees, no oaks, roses or dandelions, and no grass. In spite of the absence of these plant types, there was still plenty of green anyway during the Pennsylvanian Period but little other color.

How did loose sand turn into sandstone? Over the years, as more and more sand was deposited into the geosyncline, the deposit became thicker (maybe a delta deposit), and as it became thicker the individual sand particles were squeezed together and compacted due to the overlying weight of the newly-deposited sediment. Mineralized ground water for eons filtered through and around the sand grains, sometimes the mineral in solution was calcite which came out of solution (precipitated) ever so slowly around the sand grains forming a bond between the grains. Sometimes the mineral was quartz, sometimes it was just clay, and sometimes it was a combination of these. It takes a long time for sand to turn into sandstone, sometimes it never does turn into very good (or hard) sandstone.

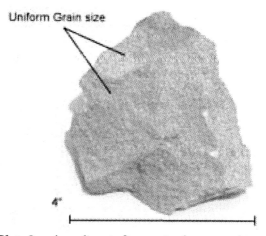

Most of the world's sandstone, and especially in the Kanawha Valley, is made up of quartz particles which were originally formed deep in the ground as hot molten magma slowly cooled. Once exposed to the weather by continental uplift, the cooled magma (granite) weathered, broke down, and dispersed the resistant quartz particles to the world. Most, if not all, of the individual quartz particles that make up the sandstone in the Kanawha Valley, and for

Plate 2 — A medium to fine-grained, gray sandstone collected from the Winifred Sandstone, Piedmont Road, Charleston. Note that the individual sand particles are clearly visible.

that matter in the world, started out as amorphous quartz (no particular crystal form) or quartz crystals in granitic rock.

The next time you go hiking, check out some of the sandstone; it is the gritty feeling stuff (Plate 2). You may find one that breaks up in your hand rather easily, or one that does not break up but the outside particles can be rubbed off with your fingers. It all depends on how well the individual sand grains are cemented, or bonded, together. Some sandstones are cemented so well that you need a rock hammer to even chip off a piece, while others crumble at the touch. Not all sandstones, though, contain sand grains all the same size. One type, worth mentioning because it is quite common in the Kanawha Valley, either as an entire bed or as a small layer inside a larger sandstone, is called a sandstone conglomerate (Plate 3).

Made up of many different size particles, the conglomerate represents a deposit resulting from a rapid deceleration of what was a fairly fast-moving, sediment-crowded stream or river. Conglomerates are deposited at the mouth of large tributaries where the fast water meets the slower water of some larger and slower-moving body of water, or meets some obstruction causing the water to slow suddenly and drop several particle sizes at once. Many of the more massive sandstone members in the valley, which will be discussed later, have distinct layers of conglomerate visible, especially toward the lower surface of the beds.

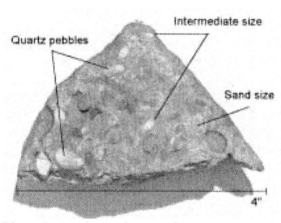

Plate 3 — A sandstone of many sizes, the conglomerate holds sand particles from fine to pebble size. A common rock in the Kanawha Valley.

The cement that bonds, or glues, the sand grains together originates from groundwater slowly percolating through the buried sand deposit. Groundwater brings with it various minerals in solution, and on occasion it also carries clay sized particles. Some of the more common minerals found dissolved in ground water are calcite, silica (quartz), and iron (oxides of iron include magnetite, hematite, limonite, and siderite). Depending upon which if any are present in the water, the mineral will slowly precipitate out onto and between the sand particles, and over time will bond the particles together forming one solid mass or layer of sandstone. In some cases, just the clay sized particles settle out of the ground water, or maybe the clay sized particles were deposited at the same time as the sand, although unlikely; in any case, they formed a weak clay or mud cement around the sand grains.

Calcite (calcium carbonate) is the relatively soft mineral seen in limestone caves as sculptured stalactites (hanging down from the cave ceiling) and stalagmites (growing up from the cave floor). It also is the main ingredient of limestone and coral reefs. Stalactites are formed when calcium rich water seeping into a limestone cave from above slowly re-deposits the calcite (dissolved limestone) in the area of the drip or flow, forming over time a calcite "icicle" or some other eye-catching shape. The point where the water dripping from above and from the tip of the stalactites hits the floor of the cave will also begin to build up with the calcite deposit, forming stalagmites. Given enough time, these two deposits will eventually "grow" and merge into one another, sometimes forming quite bazaar and beautiful shapes. An easy way to remember the difference between stalactites and stalagmites is stalactite has a "t" in it which stands for "top."

Calcite is one form of bonding agent, silica is another. Silica doesn't really dissolve in water but can exist as extremely small molecular fibers called colloids. As they become more concentrated, they will become attracted to the quartz particles already making up the sandstone, thereby "precipitating" out of the ground water as quartz, and at the same time bonding the individual particles together. Generally a sandstone that has been cemented with silica is very hard and the individual grains will not easily dislodge. Quartz is one of the hardest minerals in nature, although not near as hard as diamond.

Oxides of iron may also serve as an effective cement, especially in rather small, concentrated areas creating hard, and heavy, sometimes round or odd-shaped forms called iron concretions. Iron concretions are found in the Kanawha Valley mostly in shale layers below coal seams and are usually made of siderite. Siderite concretions fool fossil hunters all the time because of the infinite number of shapes they impersonate. Many sandstones also contain some form of iron oxide throughout the layer which when exposed will color the sandstone brown or yellowish indicating the presence of limonite, or reddish which means the presence of hematite.

Sandstone comes in many colors; for example, much of the sandstone visible along the interstate from Dunbar east to Charleston is various shades of brown to tan, in a couple of locations the sandstone has an almost orangish color. Much of the color of sandstone is influenced by the type of cement that is bonding the sand particles. Usually the shades of brown to orangish (and maybe even reddish) are caused by the presence of an iron oxide. Further east of Charleston and toward Marmet the exposed sandstone is mostly gray with dark to black vertical surface streaks (see Plate 1). In the valley a gray sandstone usually implies a calcite or silica cement. The black streaks, so obvious in some of the road cuts, are caused by ground water seeping out of the cut hillside and running down the recently exposed face of the ancient rock surface. Minerals (mostly black iron deposits) in the water will, after only a short time, discolor the ravaged and abandoned sandstone face. Weeping sandstone?

Siltstone

Siltstone consists of very fine (silt size) particles that are, for the most part, invisible to the naked eye. They can be bonded by clayey material since the silt particles may settle out of the water under similar circumstances as clay particles, although there are many examples of siltstone with other cementing agents. Siltstone is rather common in the valley and is usually found in thin layers sandwiched between sandstone and shale, and in many cases actually grades vertically into the shale becoming silty shale. The term "grade" is used when a specific rock type, say siltstone, gradually changes vertically in particle size from a siltstone at the bottom of the layer to finer clay stone at the top (or vise versa), hence changing from siltstone to silty shale (less percent silt) and then maybe to just shale (little to no silt). In other words, one type gradually changes into another type with transitional types in between over, normally, a short vertical distance. This is true for any of the primary rock types discussed so far.

When you hold a rock in your hand you hold an object of unbelievable age, but for some reason rocks just don't look old. Mummies always look ancient yet they are only a few thousand years old, a mere drop in the bucket. How exciting it would be to find a mummy 300 million years old. Imagine splitting open a big hunk of shale and finding a perfectly preserved primeval, human carcass, a being whose life mattered at the time, a being who grew up in an environment totally unlike ours today, a hot, muggy, insect-infested world with few places to hide. But during the Pennsylvanian Period there were not that many big land animals to hide from; dinosaurs would not be around for a long time yet — at least one big enough to eat you. It would be a great find, a geologist's dream for sure. But, alas, there were no humans then to donate their remains to antiquity, and would not be for another 300 million years.

There may not be any human fossils found in the rocks of the valley but the fossils that are found are just as old as the rocks holding them and just as exciting. The hills rimming the Kanawha Valley hold many ancient organisms (fossils), long dead sea creatures and plants galore. Rocks are not older than the hills, they are the hills.

Shale

Shale has been mentioned several times and its association with clay or clay sized particles. So what is shale? A good shale, that is, not a silty one, is an accumulation of clay sized particles which have been solidly compacted, feel smooth to the touch, and when looked at from the side will normally show laminations. Showing laminations means the shale consists of a series of thin layers which in most cases can be seen rather easily (Plate 4). Actually, the laminations represent thin beds of clay sized material resulting from recurrent deposits; in other words, the deposition wasn't continuous but cyclic.

When a shale is wet, more so than when dry, it may be separated, or split, along the boundary of the laminations, referred to as bedding planes. This particular characteristic makes shale an ideal rock in which to look for fossils, an attribute much loved by fossil hunters. Of all the rock types, shale is really the easiest to work with and doubtless the most productive with regard to plant fossils. Because of the fineness of the particles, the enclosed plant is not destroyed and will make its imprint into the fine material during compaction, and because of the flat nature of the bedding plane, most fossils can be retrieved intact along these bedding planes.

Plate 4 — A gray shale with iron oxide stains. Notice the thin, layered property and bedding plains, enabling the shale to be split into thin layers when wet.

If you happen to know the location of a good shale outcrop and want to do some plant fossil collecting, the best time to do it is after several days of rain, or in the rain for that matter. The moisture softens the shale and allows for easier splitting. The assembly of different plants found is a direct indication of the type of plants that grew in the vicinity during the time the shale was deposited. This is a great way to learn the different plant types and also makes for an exciting science project for natural science-minded students. In many cases, shale contains silt which results in less apparent bedding planes and usually holds less fragile fossils like leaves but may still contain more robust fossils, as described in Chapter Five.

Like other rock types, shale comes in different colors, gray to dark gray being common because of the organic content (derived from decaying plant matter) usually associated with its depositional environment. The more organic content, the darker the shale. Some freshwater shales are light tan (referred to as buff colored) but still may hold plant fossils, and some are reddish to almost purple, colored by the oxides of iron. Although not always true, a good rule of thumb is the dark to black shales commonly reflect a marine origin. Marine shales were deposited in water almost deplete of oxygen which allowed the organic material to be retained in the fine sediment and not lost as carbon dioxide. And, of course, in marine shales can be found marine animals like brachiopods and all the rest.

Look for freshwater shales immediately over coal seams, called roof shales. They commonly hold fossils of the type represented in the coal itself. In the Kanawha Valley the marine shales seem to occur wedged between some of the more massive sandstone members which will be discussed in Chapter Three. It is especially gratifying to find a marine shale because it means the sea was once at that spot — way up in the Kanawha Valley.

Along the Kanawha River most of the good shale deposits occur in relatively thin layers of several inches to around a foot or so thick. There are several much thicker. The Pittsburgh Reds Shales for one are up to one hundred feet thick in places and will be discussed in the next chapter, but these shales tend to be somewhat sterile as far as fossils go. Once shale is exposed to the weather, liberated like in a road cut, it will generally slake off, break up, and wash away much more quickly then the surrounding rock, although the more silty the shale the more weather resistant it seems to become.

Clay (Claystone) and Mudstone

Discussing clay and clay sized particles can be a little confusing unless you make it your life's work because of the different ways it can occur in outcrop. Shale has already been described as being compacted clay that will split along bedding planes. When it doesn't have laminations and doesn't split easily, but instead breaks in irregular flakes or chunks, it is called a mudstone (Plate 5). Mudstones are usually brown to dark gray and look just like what they are: compacted mud. When it has no particular breaking structure at all (crumbles any which way), and is plastic and slippery when wet, it is a clay or claystone (Plate 6). In geological terms, plastic is defined as "a property which enables a substance, such as clay, to form a pasty mass when wet, and which can be molded into a diversity of shapes, which it retains when dry."

In the Kanawha Valley there are several good, quality, thick clay deposits which were mined for bricks, earthenware, and other ceramics for many years. These mines employed many men during the late 1800s through the mid 1900s but most have now been shut down and abandoned.

Plate 5 — Mudstone, although technically just another clay deposit, has a tendency to break apart in flakes.

While fossils may be found in any of these "clay" rocks, they are easier to recover in shale because of shale's ability to split apart. Remember, all these very fine-grained rocks were deposited in slow moving or quiet water, either in a marine environment or freshwater, which allowed the finest of the fine particles to settle out and accumulate on the river or ocean bottom. I have found that the lighter-colored shales (as mentioned above) are more likely to be of freshwater origin and usually contain fossils of land plants that have fallen into a river, floated to a quiet spot, sunk, and were subsequently covered with mud.

Unless someone has found one that I don't know about, there have been no land animal fossils (except insects and spiders) found in the rocks of the Kanawha Valley. The reason is probably because during the Pennsylvanian Period

Plate 6 — Clay, or claystone, has no breaking structure. Clay in outcrop is usually light gray, sometimes mottled with red streaks of oxides of iron.

there were not that many land animals around (again, excluding the insects and spiders) except the amphibians, some lung fish, and something slowly evolving into a reptile. When these creatures died, especially on land, other scavengers and bugs more than likely scattered the remains all over the place. The find of the century would be to uncover a complete amphibian or pre-reptile fossil in some shale bank here in the valley, and if enough fossil hunters get out there and start looking, who knows?

Limestone

There is no single limestone layer of consequence outcropping within the immediately Kanawha Valley area. There are, however, several thin beds of one to two inches thick sparsely scattered about the local rock sequence. Most of these have been described in old gas and oil well logs but are extremely difficult to find in outcrop.

Limestone, by definition, is a bedded rock composed essentially of calcium carbonate (calcite) and may be of organic (animal) origin or chemical origin (Plate 7). Limestone forms in relatively shallow (less than three hundred feet deep) marine or freshwater rich in carbonates, and is usually light to dark gray in color, although I have seen red and black limestones. Limestone can be found mixed with other minerals like magnesium to form dolomite (no dolomite here in the valley), and is commonly found as silty or argillaceous limestone.

It seems peculiar to say a limestone can be formed by some organic process (made by animals) but in many cases, especially in a marine environment, that is exactly how it is made. Warm shallow seas are home to an endless number of small, if not microscopic, drifting and floating organisms. Known collectively as plankton, these aquatic organisms provide the first step in the food chain by being fed upon by bigger animals which in turn are fed upon by yet bigger animals. These minute invertebrate creatures possess tiny external shells made of calcium carbonate (calcite) which they extract from the sea water. When they die their shells slowly accumulate on the sea floor. Eventually, and possibly with the help of other calcium carbonate-secreting bottom dwellers, and because there are so many of these little animals, the sea floor becomes covered with a white, calcareous ooze of shell remains.

Depending upon how long the shallow sea remains a shallow sea, with all else being constant, will determine how thick the deposit of ooze becomes. After several million years, or who knows how long, the depositional environment will eventually change (sea level change, water temperatures change, terrestrial river influences, etc.) such that mud or sand begins to cover the limey ooze. The ooze is compacted slowly by the increasing overlying weight of more and more sediment, the water is squeezed out and the deposit will, over time, re-crystalize to become limestone.

Plate 7 — Limestone specimen from Greenbrier Limestone, Lewisburg, WV. The Greenbrier Limestone was deposited during the Mississippian age.

Water depth can fluctuate in several ways. As sediment accumulates for eons on the sea floor the sediment becomes thicker while, obviously, the water above becomes more shallow. On the other hand, the land itself may rise giving the illusion that the sea level went down. Evaporation of an inland sea or lake, in-filling of terrestrial sediments, etc. are other common causes of water level changes. Regional uplifts are quite common in the geological past; the uplift that occurred on a grand scale in the Kanawha Valley in the Permian and Triassic Periods will be discussed in Chapter Three.

An example of chemical origin of limestone, going back to the warm shallow sea, is when the surface water itself is saturated by soluble (in solution) calcium carbonate. In this case, atmospheric conditions such as pressure change or changing concentrations of carbon dioxide can cause the direct precipitation of calcite from the water as small crystals which settle to the bottom, slowly accumulating white calcareous ooze and, well, you know the rest of the story.

Nowadays, warm shallow seas, like down around the Bahamas, are the ideal environment for a host of marine animals. It was no different during the Pennsylvanian Period when brachiopods, trilobites, fish, plankton, and all the others of the day swam, crawled, floated, or just attached themselves to the sea floor and waited for the small fry to drift by, or in some cases ate each other. When a bottom dweller with a hard shell died, like the brachiopod, its shell settled into the calcareous ooze, was subsequently covered with more ooze, and later became fossilized. Most marine limestones are thus an excellent source of fossils. Freshwater limestone, on the other hand, usually lack fossils all together or are scarce at best. Although lake limestones can form under similar conditions, they lack the prolific hard-shelled bottom dwellers (with the exception of snails) of the marine environment. Usually found in outcrop as thin-layered, light gray to white, and nodular, lake limestones frequently occur in some of the red shale beds of the Kanawha Valley. Limestone does not form in a fluvial (river or stream) environment.

King Coal

Coal is considered a sedimentary rock because it forms from the accumulation of preexisting solid material; in this case, plant matter (Plate 8). What West Virginian does not know the importance of coal to the Kanawha Valley and to the whole state of West Virginia? There were hundreds of coal beds (also called coal seams or coal veins) formed in the recurring swamps of the Pennsylvanian Period, some only a fraction of a inch thick and at least one over ten feet thick — several of which outcrop in the hills around the Kanawha Valley.

The coal seams outcropping along the road cuts here in the valley were derived from the accumulation of plant matter during the Pennsylvania Period, back when, as mentioned before, vast, lowland swamps covered the area. The swamps were home to a hoard of trees, ground-creeping vines, ferns, and I suppose, weeds. When a plant died, it ultimately ended up as part of the accumulation of dead plant remains on the swamp floor. Just how long the swamp lasted would determine the thickness of the accumulated and partially decayed plant

Plate 8 — A typical blocky coal of the Kanawha Valley, often with yellow streaks which are accumulations of sulfur minerals along bedding planes. Collected from the Coalburg seam.

matter. As the plant matter built up, the bottom layers began to be compressed because of the weight of the water and additional plant remains. This compressed, partially-hardened and decayed assemblage of brown to black plant pulp is called peat, the first step in the formation of coal. Peat also forms in grassy-like marshes called peat bogs. A major plant contributor to some of today's peat bogs are high-stemmed mosses like the *Sphagnum* moss. Modern peat bogs are mined and the product (usually identified as some type of peat moss) is sold as a soil additive for plants. Peat moss has been mined for years in the bogs and moors of Great Britain and used locally for fuel.

Sooner or later the peat swamps were covered by other sediments which buried the peat deposit even further. Compactive pressure on the deposit caused by the ever increasing weight of new sediment slowly converted the peat into lignite — a soft, dark brown type of coal. The process of coal formation continues as more and more pressure and heat destroy most all that remains of the plant material except the carbon, and after several million years turns the mass into bituminous coal. The coal that the West Virginia miner risks his life for is bituminous coal which began as a raw accumulation of Pennsylvanian swamp debris. Look closely at a piece of coal, break it open and you will literally see the ghosts of ancient life; black imprints of leaves that once swayed with the breeze high above the swamp and sticks smashed almost beyond recognition, nothing left but a stack of black carbon residue that once was a part of a live, green plant.

I have heard that it takes an average of ten feet of plant debris to result in one foot of coal. If that is the case I wonder how many years of prolific swamp life it would take to accumulate the ten feet of soggy and half-rotten plant remains in the first place before it compressed into the one foot of coal? Coal in not just a black, dirty rock used to boil water for some steam generator or to be nonchalantly thrown into a furnace to keep us warm. Coal has a history, a legacy if you will, but who takes the time to look at a hunk of coal. R.I.P. — Died in the year 300,000,000 BP.

So, the next time you look at a coal seam in outcrop look at what it is — it is the chaff of thousands of individual plants all joined together by pressure, heat and time. It is a repository of prehistory, the carbon locked up in the coal was originally taken from the Pennsylvanian atmosphere, and when the coal is burned the carbon is at long last liberated just to be taken up by another plant — and maybe in the far distant future will end up as coal again.

The Pennsylvanian Period

We now know the rocks that make up the hills that border the Kanawha Valley are of Pennsylvanian age, and we now know what kind of rocks they are. But we don't yet know exactly where in the Pennsylvanian the rocks are (early Pennsylvanian, middle, etc.) and we don't yet know the local identification of the more important layers that outcrop along the highway. For instance, in Plate 1 the caption identified the black-streaked sandstone layers as the Winifred Sandstone and in Plate 8 the coal specimen was identified as Coalburg Coal. Where do these two named geologic structures, and all the rest, fit into the totality of the Pennsylvanian Period? For that, we have to start with Figure 9.

Figure 9 provides a schematic depicting the Pennsylvanian Period and how it is divided into four named subdivisions, or Series. From the oldest to the youngest (the oldest is always on the bottom), these are the Pottsville Series, the Allegheny Series, the Conemaugh Series, and the Monongahela Series, or PAC-M for short. The four parts are called series (or groups) because it refers to a specific series of rocks

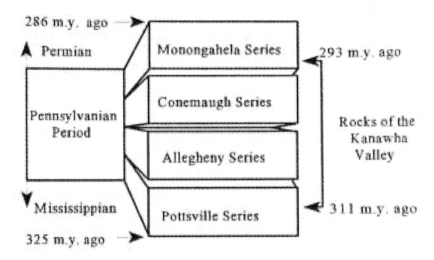

Figure 9 — A schematic showing the subdivisions of the Pennsylvanian Period. A good way to remember these subdivisions, going from the oldest (Pottsville) to the youngest (Monongahela), is by their acronym: PAC — M.

contained within each of them. No big deal, each part contains a group of studied, mapped, and named rock layers. At this point I want to introduce a new term which is usually substituted for the words "rock layers" or "rock beds." The word is "strata." This word is used when describing a number of different layers of sedimentary rock; when describing the rocks in a specific road cut I would say, "The strata is made up of a series of dark shales and fine-grained sandstones." It seems much easier to use strata than always saying rock layers.

Notice from Figure 9 only a portion of Pennsylvanian age deposits are represented within the defined scope of this book: the upper or youngest part of the Pottsville, all the Allegheny, all the Conemaugh, and the lower or oldest part of the Monongahela. The Winifred Sandstone *and* the Coalburg Coal, by the way, are both a part of the upper Pottsville Series, both of which outcrop in eastern Kanawha Valley from the Elk River to Marmet.

Because of the law of gravity, sediment is usually deposited in horizontal layers in stream beds, or along the ocean bottom, or wherever. The distinct layers of the Winifred Sandstone (Plate 1) look horizontal but they are not. They are, in fact, dipping (inclined) toward the northwest at an average rate of about 80 to 90 feet per mile, sometimes a little more and sometimes a little less. So, if a one foot layer of coal was located 85 feet above the road in Plate 1, one mile to the northwest the coal would be seen outcropping somewhere down by the road, assuming the road stayed level and ran northwest.

Another example of the overall "dip" of the strata is to consider Bald Mountain, just west of Marmet. The very top of this mountain is 1309 feet above sea level and the railroad tracks running along the river at that point is around 600 feet above sea level, which makes the top of Bald Mountain roughly 709 feet above the tracks. A great hike if not for the almost impassable thicket of seven-foot-high, mutant briers one encounters at about the 1150 foot level, but once the slashing brier torture is suffered, and your clothes and skin are ripped to shreds, the view from the top is without equal.

Considering the dip of the strata, the exposed beds on top of Bald Mountain slowly fall in elevation to the northwest and go underground in the proximity of Elk River. The tan, somewhat friable (loosely cemented), rock exposed on top of Bald Mountain is the East Lynn Sandstone of the lower Allegheny Series. This is the same sandstone that makes up the middle to upper part of the big sandstone exposure along the interstate going past the Broad Street exit ramp (Plate 17). Because of the dipping strata, if you don't want to climb Bald Mountain to see the East Lynn Sandstone, you can see it as you drive past the Farmers Market area on the interstate through Charleston, although you don't have the same view.

Quickly, in review, knowing the Coalburg Coal is of upper (or late) Pottsville age and the East Lynn Sandstone is of the lower (or early) Allegheny age tells us that the Coalburg Coal is the older of the two since it was deposited some distance below and obviously before the East Lynn Sandstone.

All the strata that make up the hills of the Kanawha Valley were deposited during the Pennsylvanian Period, but because of the dip the rocks get younger the further west you go; that is, they start out in the Marmet area mainly as the upper part of the Pottsville, through Charleston they are mostly Pottsville and Allegheny with a little Conemaugh on top. By the time you get to Corridor G exit off MacCorkle Avenue the age of the rocks are Allegheny and mostly

Conemaugh to the summits, and because of the apparent lessening of the dip around South Charleston the rocks the rest of the way to Cross Lanes are all Conemaugh age. By the time you get to St. Albans and Nitro the rocks are still Conemaugh age but with the highest hills around these two cities capped with the youngest of all, the Monongahela age rocks.

If some goliath took a big knife and sliced out a large rectangular section of earth along the trace of the Kanawha River from Marmet-Malden area to the vicinity of St. Albans and Nitro, the exposed strata would look something like that depicted in Figure 10. Although the dip of the strata has been greatly exaggerated to fit the page and the horizontal and vertical distances are not the least bit to scale, Figure 10 does illustrates the general position, both above and below the river, of the four Pennsylvanian series relative to the cities along it channel. Note that the Mississippian-age strata is well under the Kanawha Valley and has no outcrop at all in Kanawha County or any of the surrounding counties.

It is interesting to note, while looking at Figure 10, that the first wells in Kanawha Valley were brine (salt water) wells drilled in the vicinity of Malden. These wells were first drilled by David and Joseph

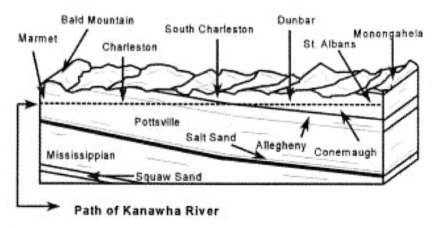

Figure 10 — A vertical slice cut along I-64 from Marmet through St. Albans showing the general configuration and outcropping position of the Pennsylvanian Period strata. Note that what can not be seen above the hills has long ago eroded away.

Ruffner in 1806 and penetrated a lower Pottsville sandstone at a depth of around 700 feet and dubbed by drillers the Salt Sand. From this lower Pottsville sandstone the salt industry was born and with it the community of the early Kanawha Valley. The much latter discovery of the Blue Creek oil field in 1911 resulted from the penetration of a lower Mississippian age sandstone at a depth of about 1850 feet identified by drillers as the Squaw Sand.

Figure 10 provides much insight with respect to the geology of Kanawha Valley. Although the strata is depicted dipping much more than it actually does, it is apparent that the Salt Sand is closer to the surface around the Marmet-Malden area. In the early 1800s, this provided easier access for the primitive drilling tools that had to be invented along the way as they drilled deeper. By the early 1900s, drilling had advanced considerably and much deeper wells were being drilled, even exceeding the depth to the Squaw Sand in the Blue Creek area. Also note that because of the more westward bend of the Kanawha River around South Charleston and because of the north-westward trend of the dipping strata, a line cut along the river would show less dip onward to the St Albans-Nitro area (as mentioned above). This is a difficult concept to see because the river actually cuts back into the dip and is sort of perpendicular to it; suffice to say, however, the strata at least has the appearance of less dip from South Charleston west to St. Albans-Nitro than it has from South Charleston east to Marmet.

One could go on about how the geology of a particular area has influenced its economic development. Maybe that's not surprising but its worth noting that many, if not most, major cities were originally settled around waterways where goods could be shipped in and out by boat, which is how the salt industry of the Kanawha Valley originated. The Kanawha River was the lifeline in and out of the valley for the early settlers since very few people during the early 1800s cared to come over the mountains, and it made shipping salt the old fashion way, by wagons back across the mountains, nearly impossible. Waterways developed as a consequence of the surface configuration of the regional geology. Surface configuration, by the way, is called topography; anyone who has used a topographic map is familiar with topography.

BOB KESSLER

The world's topography was determined by ancient land uplifts, subsidence, erosion, vulcanism, glaciers, and earthquakes, all of which are geologic processes...although man and machine have, in the last hundred years or so, contributed significantly in altering much of the topography. All the gold, silver, iron ore, diamonds, coal, oil, and gas ever found were put there by eons of geologic forces working deep underground. Where there were no waterways, towns sprang up anyway around gold and silver camps, anything worth money that could be dug from the ground became the place to be. The Kanawha Valley not only had a navigable river but had natural salt brine springs from Snow Hill to Malden, multiple layers of coal just busting at the seams, thick clay beds, building stone, and the finest hardwood timber found anywhere — all provided for by the astonishing slow pace of Nature; Nature has all the time in the world and, unlike mankind, Nature has nothing to gain and nothing to lose.

Chapter Three
Your Backyard Rocks

You made me once, maybe twice.....
scurrying about like so many mice.
The pattern is there but who knows where.
Lord don't blame me.

I wonder sometimes how the early geologists managed to do such an exceptionally good job of describing and mapping the geology of West Virginia. In the 1800s and early 1900s there were few if any road cuts to look at; the roads that did exist were mostly mud-trenched horse and buggy trails. Most areas were rugged, tree and shrub-covered, and totally inaccessible by any manner of transportation except by foot. Snakes, bears, insects, and probably mountain lions were the rule and more than likely a few mountain folk who did not take kindly to the odd-dressed stranger popping up unannounced on their side of the mountain explaining he just wanted to look at their rocks. But somehow they did it and their work and writings are still used today by the coal and aggregate industry, construction contractors, oil and gas prospectors, and people like me.

There are two types of geologists: those by name only and those who get out in the field, get dirty, and live it. My regards to the old, hard-as-nails field geologists of the turn of the century West Virginia Geologic Survey.

I have mentioned several times how flat and close to sea level West Virginia and some of the surrounding states were during the time from the middle Devonian through the Mississippian and throughout most of the Pennsylvanian. During this interval, roughly 40,000 feet of sediment was deposited in the widespread and subsiding Appalachian

Geosyncline. There is a lot of sedimentary rock under the Kanawha Valley so it's a long way down to reach the bottom, or what are sometimes called the basement rocks. Basement rocks are the ancient pre-Cambrian igneous rocks that underlie the continental deposits.

The seemingly endless erosion of the Acadian Mountains (ancestral Appalachian Mountains) and subsequent sediment deposition continued into the early part of the Permian Period where at least another thousand feet or so of sediment was deposited regionally on top of the Monongahela Series. In northeastern Kanawha County there are Permian-age rocks at the surface which continue into Jackson and Putnam Counties and become the dominate surface rocks further north on Interstate 77 toward Parkersburg. Also, Permian-age rocks outcrop north of St. Albans around the Kanawha and Putnam County line. The majority of Permian-age strata are red sandstones and shales with but a few noncommercial, thin, swamp deposits. Red-colored rocks usually indicate extreme arid conditions during the time of deposition. There were, however, no Permian rocks ever deposited within the principal area described in this book. Sometime during the latter part of the Monongahela Period, deposition in this immediate area ceased once and for all. To this day, 245 million years later, given only temporary river and lake (Teays) deposits, this area has been one of erosion.

With the help of Figure 10, it is easy to visualize how the different age strata appear in the hills as we proceed from east to west through the valley. As the lower Pennsylvanian-age strata slowly plunges deeper into the earth in a northwest direction, the upper and younger rocks begin to show on the summits.

Things were happening on a worldwide scale during the Permian; first, the old ancestral Appalachians to the east which had supplied much of the sediment for millions of years to our future home had by the early Permian been reduced to a relatively flat lowland. It was at long last over as sediment ever so slowly stopped pouring into the geosyncline. Think of the years it must have taken for mountains equal to, if not greater than, the Rockies to slowly erode to a flat plain. For countless days the sun rose over the mostly bare mountains as the rocks were imperceptibly worn away particle by particle by a thousand streams. Night storms and swollen creeks robbed the peaks of sediment and sparse foothill vegetation, all just to be washed into the lowlands below. But these sediments are not lost, they are crowded together into huge piles of sand and mud that have once again been cut up and sliced into irregular shapes by wind, water, and now, machine.

What would it have looked like on any particular day? It is absurd to even try to think of any particular day considering the time span but it must have been awesome. Geologic time is measured and discussed in alien terms; a million years here and a million years there. In our minds a million years is but an abstraction and doesn't seem like that much since it is so easy to say.

So what else was happening on a world-wide scale during the Permian? And what made the rocks of the Kanawha Valley so high above sea level today (Bald Mountain is 1,309 feet above sea level) if they were *at* or close to sea level back in the Pennsylvanian Period? And if the old Acadian Mountains eroded down to a flat plain where did the current Appalachian Mountains come from? Good questions. I mentioned earlier that plate tectonics and continental drift are beyond the scope of this book but I have to touch on these subjects just a little to answer the above questions.

For much of the Paleozoic Era, especially the latter half, our world consisted of just two super-continents (called "super" because these two landmasses contained all the exposed land in the world at that time) that looked altogether different than the continents do today. The northern super-continent, located on and to the north of the Equator, which included ancestral North America, Europe, Russia, etc., has been given the name Laurasia, and the southern super-continent, located just below the Equator, which included the ancestral South America, Africa, Australia, etc, has been named Gondwana.

During the entire Paleozoic Era these two giants were inching toward each other and ultimately destined to collide. Finally, during the late Pennsylvanian and Early Permian Periods, the collision began in earnest and continued on into the Triassic Period. As the two titans collided they formed one immense continent that we now call Pangea. Imagine, all the land in the world, excluding a few volcanic islands here and there, fused together at the seams into the one enormous land mass: Pangea.

During the collision, which no force on earth could stop, the land mass that is now northwest Africa (part of Gondwana) "slammed" into

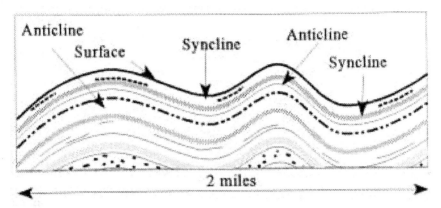

Figure 11 — Anticlines are folds in the strata, convex upward and synclines are folds in the strata, concave upward. These two kinds of folds can not be described as hills and valleys because they are not topographic features; they are underground, or subsurface, features (see Figure 12).

what is now the eastern coast of the United States (part of Laurasia), pushing and buckling the rocks high into the air along the entire length of the east coast. As the two super-continents continued to grind into each other a chain of mountains equal to the modern Himalayas was forced up, the current Appalachian Mountains were being created. The rocks closest to the collision suffered the greatest damage, what was once flat-lying strata got pushed straight up (think of Seneca Rocks), or buckled and even overturned in great mountain size folds. Upturned folds called anticlines and downturn folds called synclines were created in all sizes, from mountain-sized folds to folds several meters high, as the rocks were squeezed into a smaller space, sort of like pushing two sides of a rug together (Figure 11). Today, because of the way the highlands have been eroded, a hill, or mountain, does not necessarily imply an anticline nor does a valley imply a syncline. Depending upon how the strata was fractured or how weather resistant the top (and youngest) layers were, or even the type of drainage that existed during the time of folding can influence how the fold structure is ultimately eroded (Figure 12).

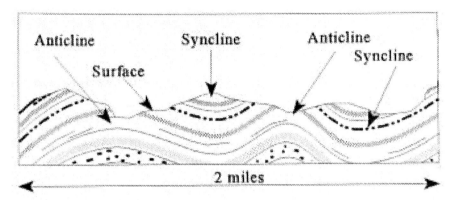

Figure 12- After weathering, note that the subsurface anticlines and synclines are still present, but now the anticlines show valleys with the oldest rocks in the middle and the synclines show hills with the youngest rocks in the middle. It can go either way depending upon the exposed rocks.

Not only did the rocks get folded and squeezed together, in many cases the thick layers of strata would fracture then slide past one another. These rock fractures where individual rock beds actually broke and moved past one another are called faults. Plate 9 is a fault showing rock movement on a relatively small scale, only a foot or so of vertical displacement. Although there is only one coal seam in the picture the "half" on the left of the fault line is higher relative to the "half" on the right of the fault line.

There are many small faults and fractures in the strata of the Kanawha Valley but it is common in the Appalachian Mountains to observe faults that have displaced the strata for tens if not hundreds of feet vertically. In most cases, the displacement occurs slowly, a little at a time over tens of thousands of years. Some movement along a fault is not vertical at all but lateral (horizontal),

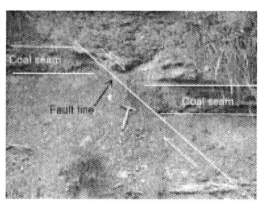

Plate 9 — A foot or so of vertical displacement of a coal seam caused by a small fault.

where the strata on either side of the fault has been displaced "sideways" relative to one another instead of up and down.

One of the more famous of this type is the San Andreas Fault of California which extends for several hundred miles along the coast while going right through downtown San Francisco. Some lateral movement along the fault occurs every so often as the two sides try to slide past each other. As the horizontal stresses build up over time (the western side of the fault is constantly trying to move north relative to the eastern side), the friction along the contact plane of the fault continues to hold each side tightly together until the stresses finally become so great that the friction gives up instantly in defeat and the western side moves violently northward.

The interval of this rapid movement and our perception of it, of course, is called an earthquake and is usually reflected at the surface with destructive consequences especially in areas occupied by man's creations. Even after the initial movement has stopped, the rocks carry the vibrations for miles around like waves in water, shaking buildings, trees, and more rocks. These so-called shockwaves go down through the earth then bounce back to the surface creating another "earthquake" called an aftershock. There are occasions where the aftershock is perceived to be worse than the initial shock, this bouncing around of shock waves will last until the energy of the earthquake is absorbed into the strata and all will again become quiet. Even when an earthquake occurs miles below the surface shockwaves, when reaching the surface, can still cause considerable structural damage. That's the problem with earthquakes, they can occur someplace else and depending upon how much movement actually occurs along the fault line, vertically or horizontally, they can be felt hundreds of miles away.

When traveling in the eastern counties of West Virginia, notice the slope, or dip, of the strata, what used to be flat-lying rock layers originally deposited at or close to sea level now are dipping (inclined) every which way and at any number of angles, and now positioned high above the sea (they were pushed up there by Africa, remember?). Further west from the "front line" of the collision a large continental region of strata (the Appalachian Plateau) was just pushed up causing the once flat-lying sea level rocks of the Appalachian Geosyncline to be elevated some distance above the sea but exhibiting only minor warping and inclination accompanied by small localized faulting (Plate 9).

The Appalachian Mountains as we know and enjoy them today are only a remnant of their once majestic size, but even today after millions of years of weathering abuse they are still the second largest mountain system in North America, second only to the Rocky Mountains, which are, by the way, much younger, having been raised in the late Mesozoic and early Cenozoic Periods (see Table 1). We will see how well the youngster fares after 200 million years of rain, ice, wind, and (of late) people.

The Appalachian Mountains are divided into two major elevated provinces: The Valley and Ridge Province which makes up, say, the eastern one-third of West Virginia (all those rocks with high angle dips) which includes the Blue Ridge Mountains (the tallest mountains in the Appalachian System), the beautiful Shenandoah Valley, the Great Smoky Mountains, and the Allegheny Mountains, just to mention a few well-known geologic structures that are not necessarily in West Virginia. The second major province and making up the western two-thirds of West Virginia is the Appalachian Plateau Province which consists of relatively flat to gently sloping strata. The "line" separating much of these two huge geologic structures is referred to as the Allegheny Front.

Since the Kanawha Valley lies within the western two-thirds of West Virginia, it lies within the Appalachian Plateau Province. All that sand and mud, coal and silt deposited so long ago in the Appalachian Geosyncline was heaved up several hundred feet above sea level which, of course, ultimately resulted in a high, vast, westward-tilting plateau stretching from the Allegheny Front through West Virginia and into the present Ohio Valley area. A long period of erosion followed which reduced the elevated plateau once again to base level — a low, nearly-featureless, gently-undulating land surface geologists call a peneplain. The rivers that crossed the peneplain were shallow, low-energy waterways that sluggishly meandered aimlessly toward the sea. When I mentioned the peneplain was "featureless" I was referring to the land surface (topography), not the vegetation. The land may have been void of any sign of a hill or discernable valley but was, for the most part, covered with dense forests and a wide range of understory plants. After being reduced to a peneplain, renewed uplift once again raised the plateau. This cycle of uplift and erosion actually occurred more than once and today we live in a period of peneplain erosion.

In short, that is how the Kanawha Valley and the rocks contained therein got to where they are now. Our valley and surrounding hills are a product of regional upheaval accompanied by chronic erosion. The rocks so spectacularly revealed in the road cuts were deposited and buried deep in the ground when time had no meaning for any of us; nature then shoved them back up, including the plants and animals that were buried with them, for all to see.

When looking from one of the high hills that line the valley, notice that all the highest summits appear to be about the same elevation (Plate 10). Interesting enough, these summits reveal to an extent the old surface of the last uplifted peneplain, the valleys in between having been formed by eons of water runoff. The rivers and tributaries running off the hillsides are once again trying to reduce everything back to sea level, no matter how long it takes. The Appalachian Mountains are the oldest mountains in North America, and, so the story goes, in the 16th Century the Spanish explorer Hernando DeSoto named the Appalachian Mountains after a Native American people of northern Florida, the Apalachee (one "p").

As the uplift of the Appalachian Plateau slowly began during the late Permian, deposition of sediment all but stopped, and renewed erosion began to occur. It occurred because the land was slowly being lifted higher above sea level. Surely during the initial phase of the uplift some deposition occurred from shallow, widespread Permian rivers from what little was left of the Acadian Mountains and some other highlands to the west, as evidenced by the Permian deposits north of the Kanawha Valley. And, with the exception of some river alluvial deposits (recent stream deposits consisting of gravel and sand), by the late Permian and early Triassic Period West Virginia and the Kanawha Valley had about all the sand silt, and mud they would ever get washed in from far away...to this day. After the Permian, the sea water would never again invade West Virginia the way it had done many times over the course of millions of years. As the flat Kanawha Valley (Kanawha Plateau?) was slowly shoved upward by the subsurface and lateral forces caused by the continental collision to the east at the end of the

Permian, the wide, meandering, and old age rivers began to merge into more identifiable rivers and at the same time began to cut down through the old sediment with renewed vigor.

The term "old age" river does not necessarily refer to a river that has been around for a long time, although most rivers have been around a long time. An old aged river is one that has done its erosion work and now zigzags across a wide, flat valley or peneplain with very little energy. In contrast, a "young" river is one that is still cutting down through and eroding the rocks, and these rivers, or streams, have steep "V" shaped valleys. The rivers snaking across the old Appalachian Geosyncline could be called old age rivers but when uplift occurred in the late Permian and these rivers got renewed energy because of the increased gradient to the sea they began to cut into the bedrock, and (as the terminology goes) they became young rivers again.

Plate 10 — The flat Tertiary age plateau can be easily seen along the distant skyline from 700 feet above the river on Bald Mountain at Marmet (looking toward Charleston). Almost as flat as a board, the plateau was shoved up while the ancient Teays River chewed away at the strata and carved the wide and picturesque Kanawha Valley. *(Photo taken by Ray Lewis.)*

After millions of years of carrying and depositing sediment into the vast Appalachian Geosyncline, now, because of the regional uplift, the plateau rivers began to slice back into the old sediment, carrying the sediment to a new resting place further down the line. West Virginia and the Kanawha Valley has been losing its share of rocks slowly ever since.

As the Kanawha Plateau slowly gained elevation the river cut deeper, and as the river cut deeper tributaries running from the plateau with ever more energy carved the plateau into a thousand spectacular hills. Today in the Kanawha Valley these summits have been timbered two or three times, punctured throughout, bulldozed, paved, and turned into business and housing developments with a few big sandstone rocks strategically placed here and there to make the site look more "natural" and then sparsely replanted with diminutive non-native, ornamental trees. How can a hilltop cut to pieces look natural?

The uplifted and tilted plateau was here first, then the valley was slowly created after trillions of tons of sediment had been eroded away. The Kanawha River has swung back and forth for eons carving into the hills creating the wide, flat flood plain we now call home, but, in truth, it was not really the Kanawha River, or the New River for that matter (in name only), as we know them today that did most of the work.

The River

As mentioned, the Kanawha River has been here a long time, actually it has been here longer than most realize. The Kanawha River, from Gauley Bridge to the St. Albans-Nitro area is just a small section of a huge prehistoric river complex called the Teays River System. The Teays River was named after a valley farmer on whose land the Teays River sediments were first studied. Some scientists believe the Teays River system carried debris from the ancient Acadian Mountains and dumped it across the low Appalachian Geosyncline 320 million years ago, all the way back to the beginning of the Pennsylvanian Period. Others say the Teays was formed 60 million years ago during the early Tertiary Period, just a few million years after the dinosaurs died out. And some say the Teays could only be 3 million years old, formed after a regional uplift toward the end of the Tertiary Period. This is a tremendous spread of potential dates, 3 million to 320 million years. One thing is certain, the Teays River cut down through the bedrock while the plateau was being uplifted, otherwise it would not be flowing north and could not have cut perpendicularly through already formed mountains. For this reason it is reasonable to conclude that the river existed prior to the last major plateau uplift which occurred slowly during the middle and late Tertiary period, giving the Teays an age of between, say, 40 and 60 million years. By comparison, the Ohio River is only about 15 thousand years old, formed after the last great ice sheet blocked the Teays River.

After the initial uplift of the Appalachian Mountains in the late Permian by the collision of the super-continents, the mountains were once again reduced to base level in the middle Triassic Period. By the late Triassic and into the Jurassic the mountains, as well as the plateau, were again uplifted. These raised mountains were again reduced to Appalachian Hills during the Jurassic and into the Cretaceous Period. Nothing much noteworthy happened in our area other than more erosion with slight plateau uplifts from the early Cretaceous to the latter half of the Tertiary when, yet again, the Appalachian Plateau was raised for the last time, culminating about 3 million years ago. Hence, the uncertainty in many minds the age of the Teays River. It is difficult to see beyond the last uplift but since uplifts don't occur overnight it seems logical to assume the ancient Teays River flowed sluggishly northward over the early Tertiary peneplain *at least* 40 million years ago.

What we do know: at the time of the last regional uplift during the latter part of the Tertiary, the Teays River existed and began to carve out its valley. With its source in the Blue Ridge Mountains of North Carolina, the Teays River flowed through Virginia and into West Virginia, and carved out the deep canyons and gorges through which the New River now flows. Continuing its trek northward across West Virginia, the Teays flowed through what is now Marmet, Charleston, South Charleston, Dunbar, Nitro, and St. Albans, the same channel through which the Kanawha River now flows (Plate 10). At Nitro the Teays didn't turn northwest, though, as the Kanawha now does, flowing northward toward Point Pleasant; the Teays instead continued straight west to Huntington, all the while bisecting the Tertiary plateau and widening its channel, ultimately shaping what we know today as the Teays Valley. This great river continued on through Portsmouth, Ohio, turned west and eventually emptied into a shallow extension of the Gulf of Mexico at St. Lewis, Missouri, a journey of a thousand miles from its headwaters in the Blue Ridge to its mouth at St. Lewis. At the time there was no Ohio River, it was the Teays River that drained most of Ohio and West Virginia.

BOB KESSLER

Less than a hundred thousand years ago the western half of the Teays River, which in some areas was two miles wide, was slowly overtaken by one of the last Pleistocene ice advances when it was overrun and stopped dead in its tracks around Chillicothe, Ohio. The ice sheet crossed over and blocked the westward flow of the river causing the river water to back up from Chillicothe through the Kanawha Valley, a lake of huge proportions estimated to have covered some seven thousand square miles (Lake Erie covers an area of about ten thousand square miles). During the life of the lake, which has been estimated to be around 25,000 years, large quantities of alternating colors of very fine lake sediments (clay) were deposited, the darker layers believed to be winter deposits and the lighter colors summer deposits. The alternating dark to light-colored clay layers are called varves, or in other words, this type of clay is called varved clay, and it is said that by counting the individual sets of colors (varves) the lake's age can be estimated.

There are still plenty of varved clay deposits remaining in the abandoned Teays River Valley between Nitro and Huntington; good exposures may be seen in some of the old cuts along the railroad. Some of the finest I have seen came from downtown Milton during construction of a new fast food restaurant. The clays found in this region are not just dirt to move out of the way to make room for a shopping mall, but are a living prehistorical reminder of a magnificently blue, living lake full of record size game fish and lake birds that stretched all the way from Chillicothe, Ohio to Gauley Bridge.

As the lake reached its full potential it began to spill over along the ice margin looking for a new avenue to the sea. The northern Ohio waters that used to drain into the Teays also had to find new paths. And new paths they found, when the glacial ice was finally gone for the last time. The river system is as we see it today, the Ohio River mirrors somewhat the ice sheets margins (but was actually not up against the margin) and the Kanawha River turns northwest toward Point Pleasant. The great Teays Lake was drained and would forever remain dry from Nitro to Huntington. The glaciers changed in many ways the Teays River channel from Nitro westward but today, although we now call the river by different names — the New and Kanawha, it is still the same old river and still flowing in its same old channel from the Blue Ridge to Nitro but restrained by a multitude of dams along the way.

Many years ago the Teays Valley was a great place for mammoths, mastodons, buffalo, and their tormentors the saber tooth cat. Many years ago the Teays Valley was a great hunting, gathering, and settlement site for Native Americans, and until recently the Teays Valley was a great place to farm, with its gently rolling topography and fertile soils. When you hear talk about how old the New River is, the Kanawha River is just as old, at least to Nitro, since they are both one and the same. And, in spite of the fact that there is no river there anymore, the old, dry Teays Valley from Nitro to Huntington is equally as old.

I have spent a couple of pages on the Teays River system because the old Teays River was the river that created the valley in which people now live. It was the Teays that cut down through the ancient Pennsylvanian-age strata we now call the Kanawha Valley, it was the Teays that cut the valleys and gorges from Gauley Bridge back to the Blue Ridge, and it was the Teays that cut the wide, beautiful, and abandoned Teays Valley from Nitro to Huntington. There was a time when West Virginia held one of the largest and most beautiful lakes in the world, how ironic is it that today there are no natural lakes in the whole state.

The Parts

Figure 9 shows the major parts, or series, of the Pennsylvanian Period, which are, from oldest to youngest, the Pottsville, Allegheny, Conemaugh, and Monongahela. Figure 10 identifies these parts relative to a slice of real estate through and along the Kanawha River from around Marmet to the vicinity of St. Albans and Nitro. From Figure 10, one can make several obvious conclusions with regard to the age and identity of the outcropping strata as it appears in the valley. First, there are no Pottsville-age rocks outcropping in and around Dunbar; second, Pottsville-age rocks are the predominant strata between Marmet and Charleston; third, Pottsville rocks dip below the surface between Charleston and South Charleston; fourth, all the rocks of Dunbar and

Institute are of Conemaugh age as they are in St. Albans and Nitro with few Monongahela rocks lying on the highest summits, etc.

In Figure 10 the Kanawha River was used as the line of reference through which the slice is made through the four Pennsylvanian series, but for the most part the actual boundary lines *between* the series and the location of their outcrop will generally hold true throughout the extent of the valley, within in the scope of this book, on both sides of the river and into the surrounding hills. Because of the north-westward dip of the strata, the towns in that direction, such as Hurricane, Poca, Winfield, Scotts Depot, Plymouth, and Teays Valley would contain only Monongahela-age strata as the Conemaugh (and all the rest) plunges below the surface. The economically important Pittsburgh Coal, which will be discussed later, is the lowest member (bed or rock layer) of the Monongahela Series. This was the coal that put the towns of Raymond, Plymouth, Black Betsy, and Bancroft on the map.

If the reader lives along the corridor delineated in Figure 10, he or she should by now know just about how old the rocks are under their house and outcropping in the hills around them and to what series of the Pennsylvanian Period they belong (at least approximately). The rest of this chapter will be devoted to tracing and identifying some of the more important beds and coal seams in each series that have been named and that have played such an important part in the settlement and prosperity of the Kanawha Valley. With the forthcoming information it may be possible to discern almost exactly where you are geologically.

Bed Time

There is an interesting branch of geology called stratigraphy (from which the term strata was derived) which is the study of the position, sequence, and continuity of the rock layers as they exist from one location to another. Stratigraphy also involves the study of the characteristics and attributes of rocks, like their mode of origin and their geologic history.

Most of our discussion up to this point has already been a lesson in stratigraphy. The early geologists mentioned in the beginning of this chapter did the initial stratigraphic work in the Kanawha Valley by identifying the physical and structural features of specific and identifiable layers. For example, when describing a conspicuous sandstone layer they would record, "The sandstone is coarse grained, gray, and carries pebbles (a conglomerate) toward the bottom of the layer. This sandstone always occurs 40 to 60 feet below a distinct and consistent black flint layer." With physical descriptions and approximate place in the geologic column given to each significant series of distinct rock layers, the beds can be traced throughout the valley by first locating the key bed; in the example above, the flint. A key bed could just as well be any other particularly noticeable rock layer like a thick, blocky coal seam or a consistently occurring conglomerate, etc. Anything that would be easy to recognize in outcrop.

Individual beds can be located underground by drillers while searching for known oil, gas, or salt sands. Depth measurements and rock descriptions are continuously being taken by drillers as the bit grinds its way down through the subsurface strata. By observing the material washed from the drill hole the rock type into which the drill is currently penetrating can be determined. By observing the speed of penetration (the drill penetrates coal and shale rather quickly but slows when it hits a medium to hard sandstone or limestone), an estimate of the hardness of the rock layer can be determined — all the while measuring the depth of the drill enabling the layer thickness to be judged and its depth from the surface to be recorded. When a key bed is hit, say six feet of hard black flint, the identification is easy as the drilling slows and the wash water coming from the hole turns black with hard flakes of flint. Knowing now the depth of the flint and knowing from a hundred other wells how far below the flint the oil sand is located, the drillers know where they are.

With the information gained from hundreds if not thousands of oil, gas, and old salt well logs drilled throughout the Kanawha County, and countless hours of field surveys, geologic maps have been drawn showing just where an important coal seam will appear, and will further show its position relative to the surface when located underground. A geologic map records the age, bed configuration, and subsurface structures (anticlines, synclines, dips, etc.) of strata in specific regions. This information is used by drillers when looking for oil before they drill to determine how far down they have to go to get to it.

Although simplified, Figure 13 delineates three wells drilled to a depth of 100 feet one mile apart from one another forming a straight line. Note that the horizontal scale (two miles) is greatly exaggerated compared to the vertical scale (one hundred feet). In this example, assume the drillers didn't know where they were geologically (the first wells ever drilled in the area) but just recorded the different type rocks each encountered as they drilled to the one-hundred-foot mark. The blank areas in the wells were left out to prevent too much clutter. Notice in well #1 the flint was penetrated fairly close to the surface with a coal seam almost immediately below it. Further down, a massive limestone was penetrated with yet another coal seam several feet below, and at around 65 feet into the well the oil sand was penetrated with yet a third coal seam showing up around 95 feet.

In wells #2 and #3 the flint gets progressively deeper toward the east along with the other identified beds including the oils sand. The drillers working at well #3 did not even drill all the way through the oil sand

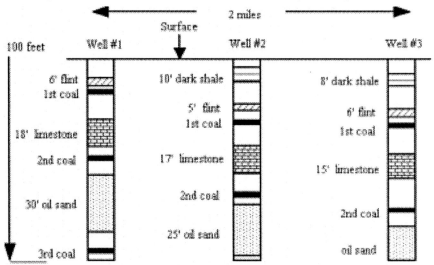

Figure 13 — Three wells drilled in a straight line through 100 feet of rock. Notice the limestone, especially, is getting thinner toward the east. Also, the drillers on well #1 reached the oil sand closer to the surface than did the drillers on well #2 and #3.

before they hit 100 feet.

To form a clearer picture of the subsurface geology, lines are drawn to connect the top of each distinct bed between the three wells (Figure 14). In so doing, Figure 14 now quite clearly shows the strata between the three wells is dipping toward the east. By knowing the distance between the wells and the depth to a certain bed the average angle of dip, or plunge, can be calculated. Had the sequence of beds been higher (closer to the surface) in well #3 than in well #2 a syncline would be evident, had well #1 shown the sequence to be lower than in well #2 an anticline would appear. At any rate, when considering hundreds of well records and actual rock outcrops, a geologic map can be assembled — not only can they show the age of the surface rocks, but they can show the estimated depth relative to the surface of key beds. Figure 13 and 14 feature only two dimensional views, add to this randomly drilled wells from all over the county and a rather clear three dimensional picture of what the subsurface rock structure is like begins to emerge.

There was a commercial on TV a decade or two ago where a famous

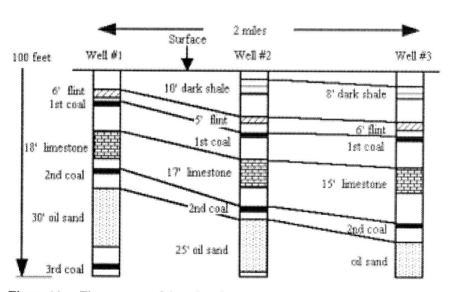

Figure 14 — The structure of the subsurface geology can be determined by drawing a line between like members in the different wells. If east is toward the right, we can say that the strata is dipping toward the east at a rate of about 15 feet per mile.

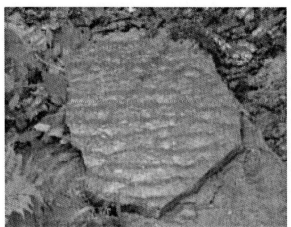

Plate 11 — Ancient ripple marks captured in a Pottsville-age sandstone. Picture taken at the Kanawha State Forest just after a summer evening shower.

athlete at the time said, "Fame is a fleeting thing." He was right, because I don't remember who he was; I don't even remember what he was selling, but the phrase stuck in my mind. Not only is fame a fleeting thing but so are mountains, only in slow motion. We have all heard our parents or grandparents say they remember when this was there or that was down by the river when they were growing up, but no one ever said, "I remember when I was a kid there was a 15 thousand foot high mountain range over behind the post office." An absurdity at best, but if we lived long enough we could say just that because the continents have had many mountains that have come and gone and come again.

We live in a time of extraordinary beauty; the mountains, valleys, rivers, and vegetation of "this time" are *our* endowments from Nature, and the only ones we are going to see in real time. Notwithstanding, the beauty of the prehistorical world is still locked up in the rocks around us and one we can nevertheless appreciate. Old stream channels filled with graded sediment, 300 million-year-old ripple marks where the shallow water of some nameless river lapped upon an ancient shore line (Plate 11), and a wonderful display of plant remains, some perfectly preserved, reveal what once was (Plate 12).

Regrettably, there is no blue preserved from the sky and water, no smells, no ambience, and no green. But who can deny this relic of yesteryear (Plate 12) found three hundred feet below the summit, squashed inside two feet of Pottsville mud, consisting only of the carbon sucked out of the Pottsville air, was real. Some may argue about its age and some may argue about its significance, but no one can argue about its authenticity. How this leaf segment got in the mud is a matter of speculation, and really of no consequence since it did, and was subsequently buried.

The only facts we can state about this partial fern is that it obviously grew, therefore it was real. It had to be part of a bigger plant connected to the ground; it was subjected to the sun enabling it to grow to the size it is, or was, and it developed from some form of fern reproductive processes (it certainly wasn't the first fern since there never was a "first" fern). And lastly, the fern got ripped from its surroundings by some force greater than the force holding it to the mother plant and ended up covered with Pottsville-age mud, where it remained for 300 million years. Some lucky outcrop walker, in this case me, then found it by halving a random slab of shale. Even without the green, I find the wonder and absolute beauty of the fern shown in Plate 12 to be a real time capsule, a black and white picture lost and then found in a haystack of rocks. The thin black residue still remaining intact on the shale is pure carbon that would burn as readily as coal because it is coal, only representing one small leaf instead of an accumulation of millions.

Plate 12 — The "tip" end of an extinct seed fern frond showing beautiful detail of the leaf development, stem, and vein structure. Collected from Pottsville-age shale in Kanawha Valley just below the Stockton-Lewiston Coal. About four inches long.

Monongahela Series

For the rest of this chapter some of the more important and identifiable layers of the four series will be discussed along with their approximate outcropping locations relative to the cities in the valley. The first, and youngest of the series is the Monongahela. As shown in Chapter Two, Figure 10, there are very few of the Monongahela rocks exposed within the scope of this book, that being on the higher hills in and around Nitro and St Albans; although, in other locations in West Virginia, the Monongahela Series ranges from around 260 to over 400 feet thick and becomes the dominate surface strata from the Kanawha River (Nitro) westward through Teays Valley and beyond. In the immediate area, most of the Monongahela has been eroded away eons ago.

What is shown in the listing below is only the extreme lowest, or oldest, units of the Monongahela Series because that's all there is with regard to our interest. All the listings given below for each series and in their arranged order were taken for the most part from the (then) West Virginia Geologic Survey's publication *Kanawha County* (1914), with some slight and humble modifications to reflect, in part, the average layer thicknesses that may exist in the valley as opposed to individual site measurements. Also, many of the identified rock units were omitted from the listings because they are either absent or poorly developed here. For a complete listing see the above mentioned publication. The major identified members are in bold face type and the significant coal beds in bold face *and* italicized type. The exposed strata of the Monongahela is given in the following listing:

Monongahela Series In Kanawha Valley

NAME (TYPE) AND DESCRIPTION	FEET THICK
Pittsburgh Sandstone (Pomeroy), coarse grained, grayish brown, massive, often holds pebbles (conglomeritic)	45 to 80
Shales, slatey	2 to 3
Pittsburgh Coal	4 to 5*

*Only represents thicknesses measured in Jefferson District which includes St Albans and Nitro area.

WEEPING SANDSTONE

The base of the Pittsburgh Sandstone of the Monongahela Series outcrops on the highest hills just north to northeast of Nitro at an elevation of around 975 feet, or about 75 feet from most of the highest summits, this sandstone actually caps several of the hills on the Nitro side of the river. The river elevation at Nitro is around 525 feet above sea level, which makes the base of the Pittsburgh Sandstone some 450 feet above the valley floor. Several feet below the base of the Pittsburgh Sandstone is the Pittsburgh Coal seam which is the only coal of any significance in the area; this coal has been mined extensively further to the north where it is better developed (thicker) in the area of Poca, Plymouth, Bancroft, and Raymond, as already mentioned in Chapter One. If you live in Nitro or the surrounding area take a drive, or better yet take a hike, to one of the summits — if you do come across a coal outcrop you will know where you are (Pittsburgh Coal of the lower Monongahela Series of the Pennsylvanian Period of the Paleozoic Era; the Pittsburgh Coal represents a deposit of about 294 million years ago).

On the south side of the Kanawha River, the Pittsburgh Coal outcrops only on one or two of the highest hills a mile or so south of St. Albans at a elevation of around 1000 feet (because of the northwest dip of the strata the coal is higher on the St. Albans side than on the Nitro side of the river). Unless you live on the summits in the locale being discussed, you do not live in or on Monongahela-age rocks but in the older and lower Conemaugh-age rocks (below the Pittsburgh Coal). If you by chance live on the broad, flat floodplain of the Kanawha River, where the bulk of Nitro and much of St. Albans are located, you live on recent river deposits called alluvium which is mud, sand, and gravel that was deposited in more recent times, say the last couple hundred thousand years or so.

Below the floodplain alluvium, maybe 50 to 70 feet, the bedrock is still Conemaugh age, and if some rocks happen to outcrop in your back yard they are, once again, Conemaugh age, between 302 to 296 million years old. By observing the rock type, color, thickness, and maybe vertical distance below the Pittsburgh Coal, or vertical distance down from the summit, the approximate formation may be determined (see Conemaugh listing below).

Conemaugh Series

Immediately underlying the Monongahela Series is the Conemaugh Series, which has been discussed several times. This means that directly below the Pittsburgh Coal (the basal member of the Monongahela Series) is the top or highest (youngest) member of the Conemaugh Series. The Conemaugh Series in the Kanawha Valley ranges from 550 to 650 feet thick depending upon where it is measured. In the St. Albans area it is roughly 570 feet thick, 150 feet of this is below drainage (below the elevation of the river).

In the easterly direction along the river toward Cross Lanes, Institute, and into South Charleston and North Charleston there is little noticeable change in the exposed Conemaugh rocks because of the almost flat-lying nature of the strata which is dipping into and across the river in this area, giving it a flat or horizontal look as you go by, but if one looked closely and actually measured the altitude of a particular exposed bed on the south side of the river it would be on average a few feet higher than on the north side, thus dipping into and across the river. Most of the primary east-west roads in the valley begin to cut back into the dip around Institute (going east) where the river takes a more southeasterly route, which then begins to expose older Conemaugh strata while the younger Conemaugh strata begins to disappear from the summits. Because of this trend (older Conemaugh strata raising above the elevation of the river while younger Conemaugh strata disappears from the top of the hills), there is no place in the valley where the entire Conemaugh Series is exposed all at once. As the lower beds reach the surface (going east), the upper beds, even on the highest hills, are long since gone, likely eroded away prior to the last regional plateau uplift. A general description of the exposed strata of the Conemaugh Series in the Kanawha Valley is given below:

Conemaugh Series In Kanawha Valley

NAME (TYPE) AND DESCRIPTION	FEET THICK
Clay and shale	5 to 15

WEEPING SANDSTONE

Lower Pittsburgh Sandstone, coarse grained, buff,
often holds pebbles .... 10 to 50

Clay and red shales (Little Pittsburgh Coal horizon) .... 8 to 10

Connellsville Sandstone, medium to coarse grained,
brownish .... 20 to 40

Limey and red sandy shale .... 20 to 65

Morgantown Sandstone, medium coarse grained, brownish 20 to 60

Red and limey shale .... 20 to 70

Grafton Sandstone, coarse grained, gray .... 10 to 65

Red and limey shale .... 5 to 30

Ames Limestone, marine origin, fossiliferous, thin or absent ..0 to 4

Sandy shale and sandstone .... 20 to 35

Pittsburgh Red Shales .... 30 to 100

Saltsburg Sandstone, medium coarse, gray, massive .... 20 to 40

Shale and thin sandstone .... 1 to 10

Bakerstown Coal (thin to absent in Kanawha Valley) .... 0 to 3

Shale and thin sandstone .... 0 to 30

Buffalo Sandstone, coarse grained, gray to light gray, pebbly,
often separated from the **Saltsburg Sandstone** [above]
by only a thin bed of shale .... 25 to 80

Elk Fire Clay .... 5 to 10

Mahoning Sandstone, coarse grained, grayish white, often reddish due to iron staining, often pebbly toward base, **holds petrified logs** .... 15 to 80

The above listing features all the important Conemaugh beds making up its thickness in the Kanawha Valley, and as such may be used as reference or guide for the identification of Conemaugh-age rocks in any given area. As an example, consider Ridenour Lake in Nitro, located on Blakes Creek at an elevation of around 650 feet above sea level (Elevations are readily available from USGS topographic maps). This elevation positions the lake about 250 feet below the Pittsburgh Coal. The red beds around the lake are rather thick and when it rains the streams draining into the lake carry red mud turning the lake a nice shade of red, especially in the spring. These are the unnamed red beds that lie below the Morgantown Sandstone (see Conemaugh Series listing above).

Since the Conemaugh Series was deposited between 302 and 296 million years ago, a period lasting about 6 million years, and the red shales below the Morgantown Sandstone were deposited about a third of the way through the Conemaugh, that would make the red beds surrounding and giving up their muds to Ridenour Lake about 300 million years old.

Of coarse this is not accurate to any great degree because different depositional environments take different amounts of time to build up sediment, and there may have been long periods of time during periodic uplifts when erosion instead of deposition took place, but these numbers give a sense of the approximate age of the rocks. Although many red beds occur in the previously mentioned Monongahela Series, all red beds outcropping or exposed at the surface in the valley will be of Conemaugh age since only a small portion of the Monongahela Series is exposed and there are no red beds in the lower Allegheny and Pottsville rocks.

The strata of St. Albans and Nitro and for several miles on either side were thus deposited during the Conemaugh Series of the Pennsylvanian Period with (again) only the highest summits still represented by the little that is remaining (after eons of erosion) of the Monongahela Series.

If you live anywhere on either side of the Kanawha River (within a mile or so) between the St. Albans-Nitro area and South Charleston you live in Conemaugh-age rocks, from the lowest stream bed to the highest hill. As mentioned before, because Cross Lanes is a couple of miles north of the river and at a higher elevation, many of the hilltops are capped with up to 150 feet of lower Monongahela rocks but for the most part Cross Lanes proper, Institute, Dunbar, and, of course, South Charleston are entirely located within Conemaugh-age rocks.

The Conemaugh Series continues through South Charleston (as a point of reference for both sides of the river) to within a mile of Elk River. Rock Lake Village, Roxalana, and North Charleston (all the way to the top of the hill) are all situated on and surrounded by Conemaugh age rocks, there is nothing else. As you travel from the St. Albans-Nitro area east toward South Charleston, you travel back through time some six million years from the upper and youngest members of the Conemaugh to the lowest and oldest members. There is one exception to all of the above (isn't there always?) and that is in the channel of Davis Creek about a quarter mile from its mouth on the Kanawha River.

There are two extremes to a creek: where it starts and where it ends. Where a creek, stream, or river starts is called its "headwater." For instance, the headwater of Davis Creek is on the back side of Kanawha State Forest at an elevation of around 1,300 feet beside the gravel road leading over the mountain to Hernshaw. There isn't much to Davis Creek at this point, a trickle of water mixed in with layers of forest clutter, but still worth seeing. If one were to walk downstream from the headwaters for fifteen miles following the same path as the creek, one would eventually come to the other extreme where Davis Creek empties into the Kanawha River around Jefferson Road in South Charleston at an elevation of roughly 590 feet — a vertical drop of around 710 feet or an average gradient (elevation drop) of about 47 feet per mile. The point where the creek empties into the river is called its "mouth." So, roughly a quarter of a mile back upstream from the mouth of Davis Creek and along Jefferson Road toward Corridor G is an exception to all the Conemaugh in the South Charleston area.

Although the Conemaugh continues to be the dominate age of the rocks in the Davis Creek valley, the underlying Allegheny Series rises out of Davis Creek at this point and continues rising on toward Route 119 (Corridor G) and beyond. Those familiar with this section of Jefferson Road know the elevation of the road increases as it ascends and then passes over the high hill then drops back down to almost creek level when approaching Route 119. A quarter mile from the river, Allegheny rocks begin to break the surface along the creek bed as they slowly rise in a southeast direction (the same direction as the road) but as the road goes up and over the hill the Allegheny is once again buried and is not exposed until the other side of the hill is reached at Route 119, where about 45 feet is exposed in the lower part of the roadcut. Notice the gray sandstone ledge just to your right when you enter Route 119 off Jefferson Road...that's the top of the Allegheny and the tan-colored ledge sitting above it is the basal or bottom member of the Conemaugh, the Mahoning Sandstone.

So, if you live on Jefferson Road beyond its intersection with Kanawha Turnpike to the base of the hill, you live on Allegheny-age rocks. Proceeding up the hill and most of the way down the other side, you live on Conemaugh rocks, which includes the basal member of the Conemaugh, the Mahoning Sandstone. The rocks outcropping all along Kanawha Turnpike from the South Side Bridge through Spring Hill, past the turn to Little Creek Park, past Village Drive, and all the way to its intersection with MacCorkle Avenue and beyond are Conemaugh. In this stretch, as mentioned before, Conemaugh-age rocks extend from the road elevation to the top of the hills — on both sides of the river.

Continuing up Davis Creek past Route 119 and toward its headwaters, the Allegheny continues to raise thereby pushing the Conemaugh from the hilltops. By the time you get to the headwaters of Davis Creek, and owing to the dip, even the Allegheny is gone and replaced by the underlying Pottsville, everything else has disappeared from the summits — eroded a long time ago.

From the stratigraphic sequence listing above, notice there are several red beds that were deposited during the Conemaugh, and if you have seen one red bed you have seen them all. It is difficult to identify one series of red beds from another unless you know something about the over or underlying strata, and in the Conemaugh most of the strata looks pretty much alike because it seems there is one massive sandstone after another and each separated by so many feet of red beds.

To describe the different and individual rock layers by name, the early geologists used easily identifiable rock layers like the Pittsburgh coal as a reference, or stratigraphic marker, and measured down to keep the various sandstones and shales they were studying in some kind of vertical order. Or, if the occasion arose, they measured up from a very persistent flint layer located in the Pottsville Series. Also, some sandstone formations were more easily identifiable and traceable than others because of color, grain size, thickness, type and hardness of the bonding agent, etc. As previously mentioned, the early West Virginia Geologic Survey geologists did all the work; we must now take the information provided and get out of our cars and into the field and see if we can locate the different members ourselves by using their descriptions and measurements.

If the strata dips in the northwest direction then, obviously, it rises in the southeast direction, one way you go down dip and the other way you go up dip. Just like any hill, you can go down it in one direction and up it in the other direction. We have discussed this several times but it is important to get the "feel" for the orientation of the strata in the valley. As strange as it may sound, when traveling east from Nitro through Dunbar, North Charleston, West Charleston, Malden, and through Belle, the relatively flat roadways that parallel the river (Route 60 on the north side and MacCorkle on the south) are actually going "up dip" and, all the while, going from younger to older rocks. Figure 10 shows this well.

The subsurface geologic structure of the strata has no association with the weathered hillsides or eroded and flat floodplain. Road contours generally follow the surface topography with the exception of the interstates, which have little regard for hill or valley. The hill traversed when on Goff Mountain Road, which by the way consists of Conemaugh-age strata, has nothing to do with the subsurface structure; the road just follows the old, weathered surface of Goff Mountain.

The uppermost important member of the Conemaugh Series is the Lower Pittsburgh Sandstone which lies a few feet below the Pittsburgh Coal of the Monongahela Series, and since the location of the Pittsburgh Coal has already been discussed it is clear that the Lower

Pittsburgh Sandstone of the Conemaugh (directly below the coal) would be found outcropping on higher elevations of the St. Albans area at around 900 to 950 feet above sea level. This rather massive sandstone continues to rise to the highest summits in a southeast direction until it, too, fades from the ridge tops (eroded away).

Below the Lower Pittsburgh Sandstone there occurs several feet of red shales which in some locations reach ten feet thick. They, too, take their turn rising eastward to nothingness. It's as if all the rocks below are pushing the Conemaugh rocks up and off the top of the hills the further east they lie. But if you happen to have that nasty old red clay soil in your yard or garden and can't get much to grow in it and wonder where it all came from, check out where you live with respect to the rise of the Conemaugh strata. Because of the multitude of red beds cluttering up the place in the Conemaugh Series, even the sandstone members on older exposures look reddish just from their proximity to the red beds.

Below the red beds of Ridenour Lake is the Graften Sandstone and below that are more red beds. The Grafton Sandstone, or parts thereof, can be seen outcropping in the lower elevations along 21st Street while going to or leaving Ridenour Lake.

From five to thirty feet below the Graften Sandstone occurs the horizon of the Ames Limestone. This limestone represents one of the last (if not *the* last) encroachment of the sea into the Kanawha Valley. Although poorly developed here, the Ames Limestone elsewhere carries abundant marine fossils such as brachiopods and crinoid stems (commonly referred to as sea or rock lilies) which are distant relatives of the star fish (see Figure 22, page 183), and can attain thicknesses of seven to eight feet in other localities. Because it is so poorly developed or absent altogether in the valley it is difficult to find and has been reported to be almost devoid of fossils.

Below the Ames some 20 to 35 feet is a very prominent and persistent layer of freshwater red shales known as the Pittsburgh Red Shales (why did the early geologists name everything after Pittsburgh?). In the vicinity of Dunbar these shales outcrop at around the 700 foot elevation, or about 150 feet above the river and can be seen in the road cut on the Dunbar side of the Kanawha River when crossing the I-64 bridge going west from South Charleston.

WEEPING SANDSTONE

Plate 13 — A massive exposure of the Saltsburg-Buffalo Sandstone members at Rock Lake Village, South Charleston. At this location the two sandstone members have no observable parting.

Further west of the bridge the Pittsburgh Red Shales are particularly well developed and outcrop up Dutch Hollow Road and into Wine Cellar Park. Here the reds are from eighty to one hundred feet thick extending from the road up to, and a little beyond, the elevation of Laura Anderson Lake. Curiously, alternating vertically with the red beds throughout their vertical extent are layers of a greenish argillaceous (clay) siltstone. On good exposures this makes a rather colorful display but for the most part the red beds once exposed are easily weathered and tend to crumble after going through several wet and dry cycles. Occasionally the red beds take on an almost purplish color; a very good exposure of the Pittsburgh Red Shales is south on Route 119 (Corridor G) in the road cut opposite the furthest shopping complex.

Notice from the sequence listing for the Conemaugh Series above that there is not much separating the Saltsburg Sandstone from the Buffalo Sandstone in the Kanawha Valley. A good exposure of these two sandstone members, where they occur almost one directly on top of the other, is in South Charleston at an old sandstone quarry and later a recreational lake called Rock Lake (now an amusement park, see Plate 13). Here about thirty feet of the Buffalo is exposed above the water with the darker Saltsburg above.

The Saltsburg-Buffalo Sandstone sequence was deposited in the bygone, sinking Appalachian Geosyncline somewhere around 301 million years ago. In this area there are no shale or thin coal seams separating these two massive sandstone beds.

I wonder, does anyone bother to look up at these rock cuts, does anyone care to tell their children that these ancient and natural structures are not just a staged backdrop for fun and games, but represent something that was already ancient beyond belief when the dinosaurs walked across West Virginia? Sabertooth cats, giant mastodons, and the ancient ancestor to the horse ranged over these rocks, one looking for something fresh to ambush while the others browsed the treetops and grazed the hillside respectively. This, of course, was before the rocks were all cut up.

A rock is more than just a rock, it was created grain by grain, therefore it has an age. It was formed in its own distinctive environmental setting which gave it color, thickness and internal structural features unique to itself, and its location in the valley has absolutely nothing at all to do with humans. Actually, it's the other way around; we humans have everything to do with the existing topography of the valley. We build on the hills all the while hoping they are strong enough to hold up our house, we slice right through them for our roads and get upset when a rock falls and blocks our path, and we invest in seemingly tranquil tributary valleys which drain scores of square miles of hillsides and wonder why it floods from time to time — and then in disbelief wonder who to blame.

When breaking open a rock and exposing the freshly broken surface to the sun (a simple task indeed) it's the first time in millions of years those sand grains have seen light, and yours are the first eyes in all eternity to see them. I find that awesome. This experience is especially gratifying when exposing a fossil for the first time.

Because the rocks in the Kanawha Valley, including the Saltsburg-Buffalo Sandstone sequence, were already a couple hundred million years old during the age of the dinosaurs, there are no dinosaur fossils found in them. There are no Cretaceous ancestral birds either, no saber toothed cats, no mastodons...the rocks were already here. From the other prospective, looking into the past, there are no Cambrian trilobites or Silurian jawless fish; none of the rocks in the Kanawha Valley had yet been deposited. Rocks can only contain what organisms they capture at the time of *their* formation. In Chapter Four the organisms the rocks did manage to bushwhack will be discussed.

The lowest member of the Conemaugh Series is the Mahoning Sandstone. The Mahoning member usually occurs as a medium-grained, massive sandstone, occasionally grayish white toward the top but mostly tan to brownish throughout. The individual sand grains of the Mahoning are weakly bonded with clays, some interspersed silica, and limonite. In many areas where the Mahoning has been exposed in a road cut, erosion has caused the sand grains to pile up on the side of the cut or in the ditch line. Often the Mahoning is iron stained and more resistant to weathering toward the bottom. There also seems to be a persistent conglomerate layer a few inches thick several feet from the base of the Mahoning. The tan to brownish sandstone exposed along the south side of I-64 in the vicinity of Montrose Drive and the Tech Center is the Mahoning Sandstone (Plate 14). If you live on the north side of the river, the very noticeable rock overhang on the west side of Edgewood Drive is in the Mahoning Sandstone.

Plate 14 — Although scarred by the drill and maligned by graffiti, this majestic exposure of the multi-colored and massive Mahoning Sandstone is seen by thousands of I-64 travelers a day. The Mahoning Gateway to South Charleston.

But there is more to the Mahoning Sandstone than its color or grain size; toward the bottom, or earliest deposits, of this massive sandstone there occurs something so rare that few sandstone beds in the world hold. Eight to ten feet above the base of the Mahoning Sandstone there occurs in scattered outcrops and road cuts petrified logs — honest-to-goodness, real, 302-million-year-old tree wood that has turned to solid stone (see Plate 35, Chapter Four).

The cell structure in this wood is beautifully preserved by silica and has been found in one location preserved by pyrite (fool's gold) as though the trees lived just yesterday. They did not, however, live just yesterday but 90 million years before the trees so celebrated and protected in the Petrified Forest of Arizona were seedlings (Triassic Period). Do other states know how to promote their natural resources better than us? I have seen quarry stone used for decorative house frontage and chimney construction, unknowingly including crushed petrified wood in their stock. How much of this ancient wonder of Nature has to be lost to back roads, valley fills, and housing development before at least one small site is preserved?

In Chapter Four this very unusual occurrence will be further discussed along with an attempt to explain why it occurs and the difference between real petrified wood and wood casts and impressions which are quite common in all Pennsylvanian-age rocks.

Allegheny Series

Immediately under the Conemaugh Series (under the petrified-wood-bearing Mahoning Sandstone) is the top of the Allegheny Series. The deposition age of the Allegheny Series is from 305 to 302 million years ago, or about three or four million years, and represents an accumulated sediment thickness here in the valley of between 100 to 250 feet, not real thick compared to the other series but containing at least one important coal seam that has been mined in Kanawha County for over one hundred years.

Once we start discussing some of the strata of the Allegheny Series notice there are no red beds included, which were so numerous in the Conemaugh Series. This occurrence, or lack thereof, is an important distinction between these two series. Red-colored rock layers may occur several ways; the most common is when sediments are weathered from a warm, humid area (in our case the ancient Appalachians) and deposited on well-drained slopes. In the presence of very fine hematite (iron oxide) particles, the red silts and muds accumulate and are washed down the slopes into the lower basins (troughs) as layers of red (hematite) sediment. If these deposits remain in a well-drained environment, such as a large piedmont area or restricted floodplain, and are not submerged for an extended period, they will remain red. Hematite particles, being clay sized, will tend to settle out with the finer silts and clays. Because of the settlement or depositional process, there are few red sandstones formed — the hematite particles don't settle out of suspension when the bigger and heavier sand particles settle out.

The concentration of hematite in clay deposits will determine the eventual color of the shale layer, some red beds being almost a maroon color. If, however, the red sediments are washed into or covered by a lasting water environment (swamp, ocean, etc.) of little to no "free" oxygen (called a reducing environment) the red beds will after time alter to gray or black depending upon the amount of organic material present. Using this scenario, the red shales and silts of Kanawha Valley are thus freshwater deposits (river floodplain and alluvial deposits) and should hold no marine organisms such as brachiopods, trilobites, crinoids, etc.

Another difference between the Conemaugh and the Allegheny is the lack of minable coal seams in the Conemaugh when compared to the Allegheny. Although coal swamps did exist during the Conemaugh, scattered throughout the vast Appalachian Geosyncline, for the most part it was a time of widespread, sediment-choked rivers actively depositing hundreds of feet of sandstones and shales. The Allegheny was also a time of widespread, sediment-laden rivers as evidenced by several massive sandstone deposits of that age, but these river deposits were frequently interrupted for long periods with the development of swamplands, which have provided much of the commercial coal taken from Kanawha County.

A last word about red beds and their occurrence; iron as an element (Fe) weathers from natural exposures in several different combinations of oxygen, water, and other molecules which, when deposited with other sediment, tend to color the sediment. The actual molecular structure of the iron compound will determine the final color of the deposit. For our purposes, when the exposed rock layer is reddish it contains hematite, when yellowish to brown (sometimes almost orangish) it contains limonite; and brown to dark brown commonly implies the presence of siderite. The different iron compounds are responsible for many of the beautifully colored rocks seen in the valley.

The multicolored pigments used by Native Americans for ceremonies or war paint were derived mostly from iron-stained muds, especially the red, earthy hematite known today as red ocher.

The uppermost bed of the Allegheny Series finally comes to the surface near the Patrick Street Bridge on the south side of the river and just west of Stonewall Jackson School on the north side of the river. On both sides it slowly rises southeastward and makes its appearance up the lower elevations of the tributary streams as it did along Davis Creek. But, although the Allegheny rises to the surface here and becomes ever more present the further east and southeast you go, if you live on Fort Hill, in Weberwood, Edgewood, and Louden Heights; Hillsdale, South Hills, Spring Hill, or on the runway of Yeager Airport, you still live on Conemaugh-age rocks with the Allegheny outcropping lower down from the summits. The Allegheny Series is not very thick, so it does not take long for its total extent to "move" from the valleys up the side of the hills toward Marmet and Belle where it caps some of the highest hills at over 1300 feet above sea level, or roughly 700 feet above the river. One of these high hills is Bald Mountain just northwest of Marmet which was discussed earlier when talking about the East Lynn Sandstone. The East Lynn Sandstone, as shown in the listing below, is the lowest sizable sandstone of the Allegheny Series.

Several of the coals of this series are more economically important in the northern part of West Virginia where they are of extreme commercial thickness. Listed below, from the youngest to oldest, are the principal strata of the Allegheny Series outcropping in the valley:

Allegheny Series in the Kanawha Valley

NAME (TYPE) AND DESCRIPTION	FEET THICK
Upper Freeport Coal, usually replaced by shale and clay	0 to 3
Bolivar Fireclay, gray to white, occasionally silty with red or yellow iron oxide streaks	1 to 3
Upper Freeport Sandstone, gray to buff, medium grained	50 to 100
Sandy shales, clay and coal streaks	5 to 10

Ruffner Fireclay .... 4 to 12

Sandy shale with thin sandstone layers, gray .... 5 to 20

Upper Kittanning Coal
(not well developed in Kanawha Valley) .... 1 to 4

East Lynn Sandstone, massive, gray to buff,
occasionally pebbly .... 30 to 100

Shale (slatey), gray, silty .... 2 to 5

Number 5 Block Coal (considered the Lower Kittanning
Coal), occasionally occurs as two separate seams .... 0 to 6

Thin shales and sandstone .... 3 to 10

    The uppermost bed in the Allegheny, the Upper Freeport Coal, is really of little importance here in the valley. In some locations the coal horizon (where it is supposed to be found in the geologic sequence) is absent or represented by a thin, dark, shale layer. In most cases, even the shale is absent and replaced entirely by the Bolivar Fireclay. I would not have even included the Upper Freeport Coal in the listing above if not for its importance as a major commercial coal seam in other areas of the state, especially in the northern part around Thomas and Davis, although it was mined locally in the area around Ruth (Route 119 around the shopping district) during the early 1900s where the coal was reported to be from one to three feet thick. A reasonably well developed and localized occurrence of the Upper Freeport Coal is exposed on the east side of Greenbrier Street at the traffic light at Oakridge Drive where it is about two feet thick (Plate 15).

    If the Upper Freeport coal, the highest member of the Allegheny Series, outcrops along the entrance to Oakridge Drive then anyone living up Oakridge Drive would live above the Allegheny, and above the Allegheny is the Conemaugh which extends from the top of the

Upper Freeport Coal to the top of the hill — and the reader already knows the identified sequence of the Conemaugh.

In many places the Bolivar Fireclay replaces the Upper Freeport Coal altogether (the coal is thin or absent), thus becoming the uppermost bed of the Allegheny Series. Find this particular clay layer in outcrop, then look up several feet into the exposed Mahoning Sandstone and you may find the petrified wood.

Plate 15 — A two feet thick layer of the Upper Freeport Coal outcropping on Greenbrier Street. Here the coal is separated from the Mahoning Sandstone by roughly four feet of clay and silty shale.

The Bolivar Fireclay occurs as a light gray to white deposit, silty in some locales, occasionally showing red and yellow streaks of oxidized iron (now and then an almost mottled appearance), and in at least one location carrying well-worn (rounded), loosely-cemented, white sandstone cobbles several inches in diameter. It's curious how these sandstone fragments became lodged within the clay layer since the two sizes (clay and cobbles) represent opposite ends of depositional environments — clay deposited in almost still water and large cobbles deposited in very rapid, if not turbulent, water. Possibly these cobbles fell from cliffs into the clay-depositing environment.

A "fireclay" is a clay deposit containing a sufficient quantity of silica or aluminum which renders it capable of withstanding very high temperatures without deforming, ideal for the manufacture of high temperature fire brick. Although not as thick and of less purity in the valley, the Bolivar Fireclay has been mined extensively in the northern part of the state for all sorts of ceramic interests. An old quarry on Cantley Drive just off Corridor G was in the Bolivar Fireclay.

Immediately underlying the Bolivar Fireclay is a massive sandstone named the Upper Freeport Sandstone. There are several excellent

WEEPING SANDSTONE

Plate 16 - A road cut exposure of the Upper Freeport Sandstone of the Allegheny Series and the Mahoning Sandstone of the Conemaugh Series. Fort Hill Drive (lower middle) actually runs on top of the upper Freeport Sandstone here with around seven feet of shales and clay (Bolivar) separating the two sandstones (hidden by the trees).

roadcut exposures of this sandstone, probably the most obvious is that on the south side of the I-64 bridge approaching the Oakwood exit ramp. Here it seems at least half of Fort Hill has been vertically removed. The Upper Freeport is the massive gray sandstone just below the Mahoning Sandstone (and Buffalo Sandstone which has been coated with grout to keep it from dropping fragments onto the ramp).

Another exposure of this sandstone found just around the hill from the previously mentioned site is seen every day by thousands of travelers going in either direction on Route 119 (Corridor G) on the Fort Hill side of the road just south of the intersection with MacCorkle Avenue (Plate 16). The massive, tan-colored sandstone sitting above the gray-colored Upper Freeport is, once again, the Mahoning Sandstone and above that is the Buffalo Sandstone (without the grout in this case) doing a good job holding up the lower houses on Fort Hill. Going further uphill toward Fort Scammons, you proceed through much of the Conemaugh Series which ultimately caps the summit at the old fort site exposing the red, limey shales which occur below the Grafton Sandstone of the Conemaugh Series — the Graften being eroded from the summit long ago.

Fort Scammons is the only public civil war site semi-preserved in the city of Charleston. At this old battle site, which used to command a strategic view of the valley (can't see the valley for the houses now), two future presidents served, Hayes and McKinley. Does anyone even know it is up there? After passing the sign at the bottom of the hill don't look for any further directions, just keep going up, and if you're lucky enough to find it, don't look for any signs at the old fort site either — and, of course, there is no place to park — it is also located on a dead end street. Imagine that, Fort Hill was actually named after a Civil War fort that once stood on the hill.

Another clay, called the Ruffner Fireclay, occurs below the Upper Freeport Sandstone which has been mined on the south side of the Kanawha River up South Ruffner and Porter Hollow. To those residents living up these two roads, if you know where the clay is you know where you are. Occasionally a coal occurs immediately below the Ruffner Fireclay called the Upper Kittanning Coal but is not well developed in the valley.

The East Lynn Sandstone which has been mentioned several times before from Bald Mountain to downtown Charleston occurs below the Ruffner Fireclay. We have discussed two massive sandstone members of the Allegheny, the Upper Freeport and the East Lynn Sandstone, each of which may reach thicknesses of a hundred feet but more often than not they are somewhat less than this. On occasion these two sandstone members come together with no partings (shale, coal, etc.) and form one colossal sandstone sequence; further, this sequence occasionally even includes the higher Mahoning Sandstone. When we discuss the geology of the Pottsville members, the top sandstone there is the Homewood Sandstone. It, too, occurs on occasion with little partings from the Upper Freeport, the East Lynn Sandstone, and the Mahoning, all together resulting in several hundred feet of massive sandstone. When this occurs the sandstone is referred to as the Charleston Sandstone.

Plate 17 shows part of the Charleston Sandstone as it occurs in the I-64 road cut just before the Broad Street exit. In this case the No. 5 Block Coal (although thin at this location) separates the East Lynn Sandstone from the lower Homewood Sandstone. Further on toward the top of the hill the sandstone sequence also includes the Upper Freeport, the Mahoning Sandstone, Buffalo, etc. Any structure secured on top of this monstrous sandstone sequence should not go anywhere for a long time.

WEEPING SANDSTONE

The No. 5 Block Coal (the bottom member of the Allegheny Series), occurs below the East Lynn Sandstone (Plate 17). This coal is also called the Kittanning Coal, which has been mined extensively in Kanawha County. In many places the Kittanning Coal occurs as three separate coal seams, the Upper, Middle, and Lower Kittanning — in Plate 17 and in most of the immediate Kanawha Valley only the Lower Kittanning (No. 5 Block) is developed. A good exposure of the No. 5 Block Coal is on Greenbrier Street at the Hillcrest Drive intersection. With the No. 5 Block Coal at Hillcrest Drive (bottom of Allegheny Series) and the Upper Freeport Coal (top of Allegheny Series) exposed just up the hill at Oakridge Drive (discussed above and shown in Plate 15), means the total Allegheny Series at this location is squeezed between Hillcrest and Oakridge Drive, a mere vertical distance of less than one hundred feet.

Plate 17 — Good exposure of the lower Charleston Sandstone and the No. 5 Block Coal, located just below the East Lynn Sandstone, which marks the boundary between the Allegheny Series and the Pottsville Series (Homewood Sandstone). The parting within the Homewood Sandstone marks a period of relative quiet deposition of clays and organic matter.

There are other rock layers that occur throughout the Allegheny and Conemaugh Series not listed in the geologic sequences given above because they are either absent or not well developed in the Kanawha Valley. In other parts of West Virginia, as well as in Pennsylvania, Ohio, and Kentucky, the geologic sequences would reflect the regional, or local, strata as it exists in those areas.

Pottsville Series

The Pottsville Series, as we already know, is the oldest of the four series of the Pennsylvanian Period. No one really knows how much sediment was originally deposited during the Pottsville age simply because in many areas erosion has reduced much of its exposed upper rocks, and in other areas there was just less deposition. In the Kanawha Valley, where most of the Pottsville is below the level of the river (within the scope of this book), the thickness of the Pottsville Series is somewhere around 1,500 feet depending upon where you happen to be doing the measuring. The maximum thickness exposed above the river is approximately 725 feet in the vicinity of Marmet, Malden, and the Belle area, and approximately 660 feet at Bald Mountain. We have already seen that the summit of Bold Mountain is 709 feet above the elevation of the river, which means the lower Allegheny-age member (East Lynn Sandstone) caps the summit to an extent of roughly the last 50 feet.

It took 20 million years to deposit the total thickness of the Pottsville Series (325 million years ago to 305 million years ago). The largest extent of any exposed portion is, again, at Marmet, Malden, and Belle but only represents a time span of six million years (311 million to 305 million years ago) with most of the older rocks still below ground. These old rocks slowly rise (on occasion dipping back down for a while) all the way to Hawks Nest and the New River Gorge Bridge area, the Nuttall Sandstone of the lower Pottsville (not shown in the sequence below) forming much of the steep, vertical cliffs of the gorge.

In the Kanawha Valley, because of the dip of the rocks, the thickest portion of exposed Pottsville-age rocks is, as mentioned, in the vicinity of Marmet (Figure 10, Chapter Two), and the thinnest is where the Homewood Sandstone (the uppermost member of the Pottsville) plunges underground, or comes out of the ground depending upon which way you look at it, just west of the Elk River. Considering the Pottsville's rise from the Elk River, it slowly pushes the above Allegheny and Conemaugh higher into the hilltops, each in turn gradually disappearing as the Pottsville-age rocks approach Marmet where they finally prevail and occupy all the land from river level to the highest summit. No matter where you live in Marmet, Campbells Creek, Malden, Belle, Witcher Creek, Chesapeake, Winifred, or even Gauley Bridge — you live in Pottsville-age strata.

Below are listed some of the more important named members of the exposed Pottsville Series in the valley and in the order of youngest to oldest (top to bottom). The Homewood Sandstone is the uppermost and youngest member in the Pottsville with the Cedar Grove Sandstone being about the oldest exposed rocks above the river which occur in the vicinity of Rush Creek and Marmet (there is still at least another 800 feet of Pottsville-age rocks beneath Marmet that will not be discussed but continue to rise, as already mentioned, all the way to Hawks Nest and beyond):

Pottsville Series

NAME (TYPE) AND DESCRIPTION	FEET THICK
Homewood Sandstone, massive, coarse grained, tan to buff.	50 to 100
Sandy shale, often with thin coal seams or coal streaks	0 to 30
Kanawha Black Flint, often contains marine fossils	.0 to 10
Silty dark shales, clay, thin siltstone partings	5 to 30
Stockton-Lewiston Coal	4 to 12
Dark shales and clays, some iron staining*	1 to 15

*The fossil fern shown in Plate 12 was collected from this shale

Coalburg Sandstone, coarse grained, massive, gray ..... 50 to 80

Shale, holds reddish iron concretions .... 5 to 10

Coalburg Coal, often multiple seams .... 4 to 10

Sandy shale and/or clay .... 10 to 30

Upper Winifrede Sandstone, fine to medium grained, gray. .10 to 110

Winifrede Coal (Black Band Coal), often multiple seams .. 4 to 12

Lower Winifrede Sandstone, grayish white, massive** .. 20 to 40
  **The ravaged sandstone of Plate 1 is the Upper and Lower Winifrede Sandstone*

Chilton Sandstone, gray to dove colored, fine grained, contains mica flakes. Considered part of the Lower Winifrede Sandstone when not separated by thin sandy shale beds ... 30 to 50

Chilton Coal, often impure, sometimes with thin clay layer. .. 2 to 6

Clay, silty shale .... 3 to 5

Sandstone and shale beds intermixed .... 20 to 40

Thacker Coal, good, blocky coal .... 2 to 4

Cedar Grove Sandstone, dull gray to dove colored with mica flakes, often flaggy (flag stone — splits uniformly along bedding planes into thin slabs — used for retaining walls, floors, walkways, etc.) .... 100 to 200

The old quarry (now a new subdivision) across the road from The University of Charleston was in the Homewood Sandstone and some of the overlying East Lynn Sandstone of the Allegheny Series. The massive and very light tan sandstone exposed along I-64 in Charleston immediately across from Laidley Field is the Homewood Sandstone. Further west (as already discussed) the Homewood outcrops as the lower portion of the great Charleston Sandstone as it is exposed in the road cut around Broad Street (Plate 17).

Another interesting point about the Homewood Sandstone, which has nothing to do with the geology of the Kanawha Valley, is that it's the same sandstone member, along with the underlying Coalburg Sandstone, that forms the escarpment over which the water cascades at Blackwater Falls State Park. If you visit the park, notice the name of the Coalburg Sandstone is not used on the sign but instead called the Connoquenessing Sandstone; the same sandstone, only a different name.

The sandstone members being discussed are of great lateral extent and were deposited as huge fluvial (river) deposits which consistently spread over the state. The thickness varies, of course, depending upon how close the deposit was to its source, and in this case during the Pennsylvanian Period the main source was the eastern Acadian Mountains.

The Pottsville Series holds some of the most persistent and extensively mined coal seams in West Virginia. The accessibility of many of the coal seams is limited or nonexistent altogether in the Kanawha Valley (between St Albans-Nitro and Marmet) since the Pottsville does not even come above ground until just west of the Elk River. Nonetheless, these seams were mined both commercially and for local fuel where the exposure and thicknesses were suitable. Further east and southeast, as the Pottsville rises in elevation into the hollows and hills of Paint Creek, Cabin Creek, Chelyan, Putney, Mammoth, and Cedar grove, much of its coal beds become exposed and offer up its thick masses of Pottsville-age plant debris for mining.

Separated from the Homewood Sandstone by several feet of silty shale and an occasional thin coal smudge is the most important and persistent stratigraphic marker of the Pennsylvanian Period. The Kanawha Black Flint is so obvious in outcrop or when hit by the drill that all other strata, either above or below, may be identified by their measured distance from it. Reaching a thickness of seven feet in some outcrop locations, this flint is a most interesting and enigmatic unit just by being where it is and what it is — a dark gray to black marine shale that in many locations holds marine fossils, yet for some reason has since its deposition turned to flint by the infiltration of silica. Neither the layers immediately above nor below have this unusual silica content.

Native Americans who lived and hunted on these grounds for thousands of years before the valley was usurped by our forefathers (I have mentioned our forefathers before) thought the flint was rather important too. Because the flint is so unique and persistent throughout the valley it is referred to as a stratigraphic key bed, or marker. The term stratigraphy was discussed before and one of the examples I gave was, if you knew the location of a unique and easily identified layer in the rock sequence (a key bed or stratigraphic marker), then the identity of the rock layers above and below could be easily mapped. Knowing where you are is particularly important to oil and gas well drillers, allowing them to estimate how much further they must drill or from which layer the oil or gas is coming...or in most cases, not coming.

The Kanawha Black Flint comes above drainage in the vicinity of the Elk River at an elevation of around 600 feet. Progressing east to southeast, it slowly rises to 640 feet at Porter Hollow Road (just above the railroad tracks), then 775 feet at the old quarry across from the University of Charleston; 850 feet on the hill leading to the Job Corp (Plate 18); 985 feet on the hill (K-Point) just east of Upper Donnally Road in Kanawha City; 1,020 feet on Donnally Hill (behind K-Mart); 1,210 feet on Bald Mountain, and then caps the hills at Marmet at 1,300 plus feet. This is a straight line distance of about eight miles and an elevation rise of 700 feet, or roughly 87 feet per mile.

Plate 18 — A four foot thick outcrop of the Kanawha Black Flint (a marine deposit) lying above four feet of light-colored sandy shale. The Stockton-Lewiston Coal lies below the sandy shale.

Many readers won't know where these places are and that's all right, this illustration is intended to show the rise in elevation of the strata from west to east by following the outcrop of the flint from one hill to another and measuring its altitude, and as the flint rises so do the other rock layers above and below it.

The Kanawha Black flint is an enigma of sorts, originally deposited as a marine clay (in some locations a silty clay) in the shallow, calm water of a protected embayment or small eastern blossom of a larger sea lane that extended from the Gulf of Mexico up through Tennessee, Kentucky, Ohio, and into Pennsylvania. Marine animals enjoyed the relative calm and warmth of the tropical sea water and their fossils are found today at many outcrop locations (Chapter Five). All the while, thick, dense forests lined the miles of beaches along the shore of this inland sea (at most forty miles wide covering the southern two-thirds of Kanawha County) and contributed a never-ending supply of organic matter to the shallow sea floor which turned the silt and clay deposit dark gray to black. If one were to walk along and into the shallow water of this Pottsville-age beach it could be described as walking in black muck yet full of marine life. How long does it take for six to seven feet of black organic muck to accumulate? Actually, the muck would have had to be much thicker than that to allow for the compacted and dense bed we see today.

Nothing about this scenario of shale deposition is unusual since it occurred many times and in many places in the geologic past. What *is* unusual, and the enigma, is why and how did it turn to flint? Most everyone knows what flint is, and that is, the very hard and smooth rock from which most Native American spear points and arrowheads were made. Chert is another term used to describe the same thing. Scientists have been trying for years to figure out the difference between flint and chert, right down to their molecular structure, and maybe the difference is in name only. A good rule of thumb is, if it is dark gray to black call it flint; if it is lighter and/or mottled with other colors, call it chert.

In many places the Kanawha Black Flint is not well developed. It occurs as a somewhat siliceous (some silica in it), slightly silty, "slatey" type; in other places it occurs as a somewhat low-grade, grayish and silty flint, and in still other places it occurs as a well-developed, high-quality dark gray to black flint. All in all, though, the flint is for the most part recognizable in outcrop and when drilling through it. Native Americans who lived or traversed through the valley for thousands of years, before any fur traders or settlers came around, mined the flint for their own implements and apparently traded or hauled the flint to other locations since the flint is found in the fields throughout the extent of the Kanawha Valley (all the way to Point Pleasant) as well as up and down the floodplain of the Ohio River Valley.

The extraordinary point shown in Plate 19 (crafted from a piece of high quality Kanawha Black Flint) was found in Kanawha County by my son, Joe. The point is a Palmer corner-notched spear point made in a time period archaeologists call the Archaic, and dates to about nine thousand years ago.

Plate 19 —A piece of high-quality black flint collected from the upper Campbells Creek area (left) and a perfect spear point found in Kanawha County made from the same high-quality black flint. Spear point about one inch across.

This particular point is so flawless with its symmetry, serrated edges, and perfectly notched corners that it may have been a ceremonial point — who would take the time to craft such a exquisite black flint point just to throw at a deer or bison when one of lesser excellence would suffice? And this is not a point made several hundred years ago, the point shown in Plate 19 was crafted several thousand years before the great pyramids were even in the design stage.

Egypt doesn't have a claim on ancient human history, for you may find something in your own backyard every bit as old, and every bit as remarkable as can be found anywhere in the world. Am I going so far as to compare a little spear point with the great pyramid of Cheops? Maybe not in size, but certainly in skilled craftsmanship. Big just takes more people — each thousand or so good at only one thing — no one person could ever say, "I built the great pyramid of Cheops all by myself."

Some of the best Kanawha Black Flint, by that I mean the highest quality, is found east of Charleston in the upper Campbells Creek area and the old mining community of Putney. Here the flint outcrops right along the road and is scattered all along the creek bed. If you live up Campbells Creek, especially around Cinco to Blount and on into Putney, you know where you are.

Because flint (at least good flint) has a very high concentration of silica (quartz), to turn a shale into flint you need two things; first you need the shale and then you need a supply of soluble silica, which is just quartz dissolved (so to speak) in water. Sea water, even today, contains a small percent of silica but not near enough to solidify much of anything, let alone several feet of shale. There are a few theories as to where the silica came from; one theory suggests that during the time of deposition the shallow coastal waters were home to a great number of sponges. It may be difficult to believe, but a sponge has an internal skeleton made of thousands of microscopic silica granules called siliceous spicules. In this case, when the sponges died and the soft parts decomposed, the spicules enriched the waters with dissolved silica which then slowly percolated through the shale as it was being deposited, thus turning the shale into flint. A real possibility since the fossil impressions of what some paleontologists perceive to be sponges (*Conularia*) have been found in the flint.

Another theory is that silica-rich waters running off an arid landscape into the restricted marine bay supplied the silica. In this case, the silica was released from the land rocks by chemical weathering, although I don't see how you could have an arid landscape and at the same time have lush forest surrounding the coast sufficient to supply the quantity of carbonaceous material to turn the flint black. Yet another idea is the silica may have come from hot geysers, some of which bring dissolved silica to the surface from deep underground which then ran into the sea and subsequently infiltrating the shale. A pretty good theory.

There are several other theories concerning the whereabouts of the silica that formed the Kanawha Black Flint but my own view, although I am sure others have thought of this too, is that the marine embayment was surrounded by thousands, if not millions, of *Calamities* trees. We know during the Pennsylvanian Period *Calamities* were a very prolific and dominate soft tissue tree which held a very high silica content. Although the *Calamities* tree is now extinct, its closest living relative today is the small, stilt-like, scouring rushes (also know as horsetails) which were used by our close ancestors to scour pots and pans because of their scouring powers derived from the high silica content in their leaves and stems. Considering the potential number of these trees surrounding the coastal waters, huge quantities of silica and organic matter could have been gathered from the forest floor for a thousand years and carried via ten thousand streams into the restricted and muddy sea lagoon as the *Calamities* grew, died, and grew again. As the trees withered, decayed, and fell into the waters, they provided their silica which then became concentrated in the protected embayment.

The resultant free silica was transported into the bay area creating a silica-rich water environment while at the same time clay sized particles and black organic material were gathering on the sea floor. The silica permeated throughout the clay layer, sometimes in very high concentrations, causing the layer to ultimately become a high-quality flint and in some areas only low concentrations causing the resulting shale to be just brittle and silty, all the while capturing many of the shells of the dead brachiopods, pelecypods, and other marine animals. But who knows, maybe more than one source of silica existed at the time and maybe all the theories are correct; whatever, the flint was formed and exists today (along with its fossils) as an excellent stratigraphic marker from which most other Pennsylvanian strata can be measured in Kanawha County. More about *Calamities*, what they looked like, and the fossils they left behind is in Chapter Four.

The Pottsville Series is noted for its coal seams; huge quantities of peat, in an on-again/off-again manner, were built up during this period. In between the accumulation of peat there were times of rapid sand and silt accumulation; other times were quiet which caused freshwater shale deposition, but it was the accumulation of peat during the Pottsville and the overlying Allegheny that defined West Virginia's future. It seems there was a long-term depositional rhythm occurring during the Pottsville age and into the Allegheny. Over and over again the sea encroached upon the land, depositing marine shales or

WEEPING SANDSTONE

limestones then receding once again, giving way to freshwater sandstone deposits some a hundred feet thick, and then the coal forming swamps had their turn. This cyclic habit continued for twenty million years throughout the Pottsville and Allegheny ages until finally breaking the pattern in the Conemaugh. But for a few infrequent and short-lived exceptions, after the deposition of the marine Ames Limestone of the Conemaugh, the sea never again ventured into our area.

There were still infrequent swamp deposits into and during the Conemaugh and Monongahela, the greatest accumulation of plant matter of all being the Pittsburgh Coal of the lower Monongahela which in some areas reached more than ten feet thick, and the final coal deposit of significance occurred in the overlying Permian Period (the Washington Coal). Then, as far as West Virginia was concerned, coal deposition in our part of the country was over. An interesting note: The great coal deposits of Wyoming were formed during the much younger Cretaceous Period.

Sometime during the Permian, the geosyncline at long last stopped sinking and slowly began to rise because of the great continental collision taking place on the east coast as Africa slammed into North America. Slowly, ever so slowly, the land was pushed up, all the while exposing to the elements more and more of the lost sediments of the geosyncline. The endless days of deposition did indeed end and the reign of erosion took over...and continues to this day.

A coal deposit starts out as an accumulation of organic debris in a swamp. Sandstone begins as an accumulation of loose sand, possibly along a beach, and shale originates as a build-up of mud layers anywhere fine sediment can accumulate. In many instances, as already discussed, and because of changing environments, these sediments mix together forming a variety of rock types. The observer must look closely at a rock, what one sees is how it formed. No one can see sunshine, wind, or blue skies but we can see rivers, forests, and what was once life — all things made possible by these natural forces.

Getting back to the Pottsville Series, I have already mentioned the anguished Winifrede Sandstone and its black-streaked face now exposed along Piedmont Road east of Charleston. The Winifrede Sandstone escapes forever underground in the vicinity of the Kanawha

City bridges.

Below the Winifrede Sandstone is the Winifrede Coal. Although not well developed in the immediate valley, the Winifred Coal was mined extensively in the Kanawha State Forest area in the late 1800s and early 1900s and was at the time called the Black Band Coal. A hike up Snipe Hollow behind the swimming pool exhibits a sealed mine that was opened on the Winifrede Coal a long time ago. Actually, there were many mines opened on the Winifrede Coal in the forest as well as the higher Coalburg and Stockton-Lewiston Coal. On the back side of the forest, at an elevation of roughly 1,300 feet (the headwaters of Davis Creek), even the Kanawha Black Flint was mined during the very late 1800s for its high concentration of iron. The flint was hauled by rail to the mouth of Davis Creek where a furnace had been built to separate the iron from the rock. Proving unprofitable, this venture was abandoned rather quickly. Here the flint, because of the iron (hematite) content, actually takes on a red color. The current road leading to the forest is built on the old railroad grade up Davis Creek, called the Kanawha and Coal River Railroad, which then led to the small mining community of Chilton.

Chilton had several hundred residents, several churches, and a post office, and used to occupy the area in the forest around Polly Hollow. After all the chestnut, white oak, poplar, chestnut oak, and walnut were plundered and sold for five dollars a tree, and most of the coal dug out, the timber and coal companies closed up shop and moved on, leaving the once hardworking town of Chilton high and dry. As expected, and typical of hundreds of small timber and mining towns, Chilton shriveled and died (isn't it easier to say a town died and avoid saying what happened to the families who made up the town?). Did the coal and timber companies care? I wonder if the lumberjacks and miners who broke their backs and breathed the coal dust received some kind of severance pay to tide them over until they got another job?

The town of Chilton was subsequently erased from the face of the earth by the CCC (Civilian Conservation Corp.) when in the 1930s they bulldozed and hauled Chilton away to who knows where. Where did all the families go? I think it would be nice today to have at least one or two of the old homes or maybe the post office still standing, just to look at and remember the good times launched by the Winifrede Coal.

I'm sure there are still many people around who remember the community of Chilton as it once was. Why get rid of everything?

Getting back to geology and the Pottsville Series, the Thacker Coal is of little importance in the Kanawha Valley (close to the bottom of the exposed Pottsville Series listing above). The underlying Cedar Grove Sandstone was quarried just west of Marmet at Rush Creek and then plunges under ground quickly between Rush Creek and Charleston, not to be seen again.

A book could be written on any single member of the Pennsylvanian Series. The sandstones are beautifully colored on weathered or fresh surfaces. They are of various grain size which betrays their origin; some hold fossil casts and coal smudges, at least one holds petrified wood, many hold ancient animals, and each tells a story of its creation. The shales of the Pottsville Series are both marine and freshwater, a good many of which contain fossils of organisms long since dead, both plant and animal, that once called these parts home. I love it.

Chapter Four
Fossil Trees of the Kanawha Valley

Too many ideas, too much thought,
makes you want to leave with more than you brought.

Fossils are the remains, traces, or direct evidence of prehistoric life. An animal's "remains," in this case, don't mean the fossilized gooey stuff of an organism, but rather its remineralized bones, shell, or exoskeleton. Only in very rare occurrences are internal organs preserved to the degree they can be differentiated within the fossil itself. The fossil leg bone of a dinosaur protruding from a sandstone bluff in Wyoming would be considered its remains, while the famous dinosaur footprints discovered in the shallow water of the Paluxy River near Glen Rose, Texas would be considered the dinosaur's trace.

Animals don't have to be as big as a dinosaur to leave their trace. Many small animals don't even have hard parts but have left traces of their daily activities. Although not very common in the rocks of Kanawha Valley, preserved worm holes and borings can be found in some of the finer-grained rocks. Actually, the worm "holes" themselves are not preserved, the holes get filled up with mud or silt and then preserved as undulating tube-like structures. As a group, traces of prehistoric organisms are called trace fossils; as a science, the mode of creation, interpretation, and classification of trace fossils is called ichnology.

Beside remains and traces, direct evidence of prehistoric life such as compressions, impressions, molds, and casts are considered by definition to be fossils. It's not easy to become a fossil. The first grim reality is the organism must die, and second, as already mentioned, it helps dearly to have hard parts (death isn't necessary to become a trace fossil). Of the almost infinite number of organisms that have ever lived and died on this planet, with or without hard parts, in the millions of years of prehistoric time, only a very small percentage have actually become fossils of any type.

When most animals (including insects) and plants die, especially land-loving organisms, their carcasses wind up on the ground for scavengers (big and little scavengers) to eat and scatter their hard parts about and/or just rot away. Today in the Kanawha Valley an animal or tree limb lying dead on the hillside or in the woods somewhere has virtually no chance of becoming a fossil, why would it?

To become a fossil, and the third and fourth requirements needed (besides dying and having hard parts), the organism must fall, be wind blown, or be washed into a depositional environment. That is, an environment where sediment is being continuously deposited; after which the organism must be covered with sand or mud rather quickly to prevent decay. Not only does all this have to happen in some kind of sequence (sometimes the organism is buried before it dies) but the potential fossil has to be imprisoned to such an extent that chemical processes caused by lack of oxygen, compaction, or by the infiltration of soluble minerals in the ground water over time will alter the organism (or its form) to a preservable state.

Today in the Kanawha Valley we live in a time of erosion, the hills around us are imperceptibly being washed to the sea while the river and its tributaries continue relentlessly to turn this area once again into a flat peneplain; we can't stop it. Nature makes mountains and then takes them away. Cloaked as a much needed spring shower just to nourish our tomato plants, Nature secretly carries hundreds of tons of hillside sediment into the Kanawha River, who notices? In our lifetime we will not notice, we may see here and there an area of soil washed out from a field or along a back road somewhere but think nothing of it. New road and housing construction, especially on the side or top of the hills, provides fresh quantities of sediment to the rain runoff. With each erosional gully created we have lost something forever, precious soil and ancient sediment that once was ours, gone.

Recall from Chapter One, most of the Pennsylvanian Period was a time of sediment deposition in West Virginia where several thousand feet of sand and mud were being piled up by sediment-choked, meandering rivers. Where the rivers allowed, vast swamps accumulated thick deposits of vegetable matter. Periodically, the sea advanced inland and covered the freshwater sediments with various kinds of marine sediments. If any time in history or prehistory it was easy (relatively speaking) to become a fossil this was the time, simply because the chances were greater after death that an organism would end up in the water and soon buried in sediment...a potential fossil paradise.

Fossils are a modern day show-and-tell that has against all odds survived a series of primeval random events. A picture of what once was, there is no greater anticipation than being out on some shale bank somewhere carefully separating each bedding plane looking for the perfect fossil, and no greater thrill when suddenly the anticipated fossil is right there in your hand. The excitement of finding a good fossil is further enhanced by being able to identify it.

Modes of Fossil Preservation

Plant and animal fossils preserved in the Kanawha Valley occur as (1) molds or casts, (2) compressions or impressions, (3) a remineralized (petrified) form of the original organism, and (4) traces. Excluding leaves, most plant parts like stems, trunks, roots, and even seeds (Plate 20) will be preserved as a mold or a cast.

Molds and casts are formed when a bulky plant part suddenly finds itself in a watery grave while being covered with anything from medium-grained sand to fine mud. As the sediment becomes thicker the plant part is completely covered, thereby slowing down its rate of decay. As more and more sediment accumulates, the underlying sediment containing the plant part slowly becomes compacted while squeezing out much of the water surrounding the particles.

How long the plant part exists in this state is anyone's guess, but ultimately the plant will deteriorate or dissolve due to groundwater still seeping through the mass leaving behind an empty cavity, or **mold,** of the original plant part. What occurs is no different than if you took an acorn and covered it with Plaster of Paris. After drying, the Plaster of

Plate 20 — A 300-million-year-old seed cast from an extinct seed fern now embedded in a silty shale, about 3/4" long. Found in Allegheny-age rocks. This seed belongs to the genus *Trigonocarpus*.

Paris can be split open and the seed removed, leaving a perfect **mold** of its form. The outside surface features of the acorn will be perfectly preserved on the surrounding inside walls of the mold. If the mold is then filled back up with, say, mud, allowed to dry and then reopened the result will be perfect copy of the seed but now it would be made of mud. The resultant "mud seed" would be the **cast** of the original acorn.

In nature, percolating groundwater carrying finer sediment may fill the mold and form a cast such as the fern seed shown in Plate 20. Probably the decay of the plant occurs at the same time as the infiltration of the finer sediment such that a complete cavity really never exists before it is filled up again.

Shelled animals, such as brachiopods and pelecypods (clams) are commonly preserved as molds and casts, as are animals with hard chiton-like exoskeletons like trilobites and beetles. The fossil brachiopod shown in Plate 21 is just half of a mold found in the Kanawha Black Flint, the other half is on the piece broken off (not shown).

Although the outside surface details of the organism may be preserved on the mold *and* the cast, there are no internal structures preserved. There is nothing inside the seed cast shown in Plate 20 but more dried mud, and, of course, there is nothing in the mold shown in Plate 21 because there is no "in." The hinge line pointed out in Plate 21 is the line, or point of contact, between the top shell (dorsal valve) and the bottom shell (ventral valve) along which the brachiopod opened

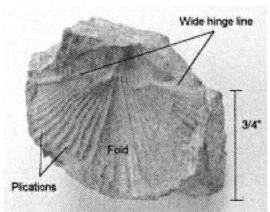

Plate 21 — A mold of a brachiopod found in the Kanawha Black Flint of the upper Pottsville Series. Note the detail of the shell plications (striations) preserved. This particular type of brachiopod, as a group, are called spirifers because of the relatively straight hinge line, deep plications, and middle fold on the bottom (ventral) shell.

and closed itself. It just so happens that the part of the mold showing is the ventral valve — or bottom shell of the brachiopod. Molds and casts are considered direct evidence of prehistoric life since there are no remains left. When discussing further the different kinds of fossils found in the Kanawha Valley there will be more illustrations of molds and casts.

Compressions and impressions are yet another mode of plant and animal preservation. In the Kanawha Valley, the majority of compressions and impressions are found to be plant parts like leaves, small sticks and stems, and some reproductive structures like cones and seeds. Found with less frequency are compressions or impressions of insects such as beetles, and virtually no shelled animals.

On page 95, Plate 12, the extinct seed fern leaf is preserved as a compression. This small tip-end of a fern frond was dislodged somehow from the rest of the tree, or maybe the whole tree fell into muddy water and only the tip-end was found; whatever, the leaf ended up in the water and was quickly buried in a silty mud grave which rapidly checked the normal decaying process. Compaction occurred and over time the original leaf material broke down chemically leaving only the black carbon remains of the plant embedded deep within the shale layer. When the piece of shale was finally split open, the surface of the bedding plane contained the compression (called the positive side) of the fossil fern, and the other side of the bedding plane — the part lifted off the compression — contained a perfect impression of the fern (called the negative side) but, as in many cases, it had little to no carbon residue.

A good example of this is seen in Plate 22. This small leaf of a true fern (a fern that reproduced by spores instead of seeds) was found in a small iron stone concretion that was split open revealing both the compression, or positive side of the fossil, and the impression, the negative side of the fossil. Although this fossil was not found in the valley, it illustrates very well the two "sides" of a fossil. Looking carefully at the positive side (left) you can see that the fossil is raised above the plane of the stone matrix and the negative side (right) is below the plane; hence, positive (compression) and negative (impression).

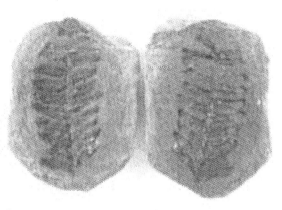

Plate 22 — A compression (left, positive side) of a Pennsylvania-age true fern found in a small iron stone concretion and the corresponding impression (right, negative side) exposed after the iron stone was split open.

The third mode of fossil preservation has many names such as altered, mineralized, re-mineralized, permineralized, silicified, and petrified. Big names that just mean about the same thing, that is, during the time span of the organism's burial minerals in the ground water either chemically replaced some the original organism's hard parts or surrounded them with new minerals — or both! When this occurs the fossil is considered the "remains" of the organism as mentioned in the fossil definition.

On many occasions, brachiopods, pelecypods and other marine fossils found in the Kanawha Valley have had their shells or other hard parts altered by soluble minerals from thousands of years of ground water seeping through and around them — without any internal soft parts retained.

Plate 23 shows a small brachiopod that is quite common in the marine shales of the Kanawha Valley, including the Kanawha Black Flint, called an orbiculoid. This small ocean dweller differs considerably in shape and body adornment from the brachiopod spirifer shown in Plate 21. Obiculoids, represented in the fossil record from the Ordovician to the present, are usually round when viewed from the top (or bottom) and somewhat conical in shape when viewed from the side. The interesting feature of this particular fossil is that it shows three different modes of preservation. First, the cast of the brachiopod is the one on the left and second, the mold, or at least half of it, is on the right. Third, and even more interesting, is the shell is still contained in the mold (the white stuff). Orbiculoid brachiopods commonly have a chitinous shell (fingernail-like material) or a combination of chitin and calcium carbonate. This shell has, since the animal's death some 300 million years ago, been altered to just calcium carbonate, a mineral discussed in Chapter Two.

Ground water, under different environmental conditions, carries different minerals in solution that have been known to replace, or re-mineralize, organisms. The best known mineral is silica because of the petrified wood of the Petrified Forest in Arizona. There, ancient conifer trees of the Triassic Period were tumbled for whatever reason, probably as a result of volcanic activity, and washed downstream in a slurry of volcanic ash, and then gathered miles from where they fell in a bath of thick mud and volcanic soup.

Plate 23 — An orbiculoid brachiopod exhumed from a gray Pottsville-age marine shale after splitting along a bedding plane. The mold (right) still contains the remains of the brachiopod's lower (ventral) shell. The small hole seen in the middle of the shell, on both the cast (left) and the mold, is where the brachiopod extended a fleshy muscle to attach itself to the sea bottom. About 1/2" wide.

Volcanic ash, having a very high silica content, enriched the ground water with soluble silica which then permeated and infiltrated even the smallest cavities deep within the woody cell structure of the entombed giants, encompassing and/or replacing each molecule chemically that was once living tissue with rock hard varieties of quartz.

There are probably thirty some minerals that have been known to remineralize the hard parts, and more rarely the soft parts, of buried organisms. But the most common by far is silica, calcite, and pyrite (fool's gold) — pyrite being the less common of the three. Plate 24 shows these three common minerals as they naturally occur in their crystalline form; that is, when given enough time and free space to "grow" into crystals. It just so happens that the minerals on this short list occur in abundance in the Kanawha Valley, although rarely to the size shown in Plate 24.

Plate 24 — Three of the most common replacement minerals found in remineralized, permineralized, and petrified hard parts. For scale the pyrite crystal is about 1/2" inch high.

The first step in identifying these minerals in the field is to know their physical characteristics, such as color, hardness, and crystal shape. These differences seem quite obvious when looking at Plate 24. The quartz crystal (SiO_2, or silicon dioxide) is, when not stained by other minerals, clear, elongated and has six flat sides (hexagon shaped) which come to a point at the top. The pyrite crystal (FeS_2, or iron sulfide) is gold colored and has a cubic form, much like a salt crystal. Pyrite is very common in sedimentary rocks and is found most any place where the rocks contain iron and sulfur that has combined chemically. Sulfur alone occurs in varying amounts in coal seams of the Kanawha Valley and shows up on older, exposed surfaces as patches of a bright yellow crust. The gold color of pyrite fooled many early gold miners into thinking they had hit it rich, only to find out to their embarrassment it was not gold at all; hence the popular term fool's gold was coined. The calcite crystal ($CaCO_3$, or calcium carbonate) is white (when pure) and the crystals, at first glance, also look cubic but closer inspection reveals a sort of "pushed to the side" cubic form called a rhombohedral.

The three minerals shown in Plate 24 represent their appearance when in crystalline form which allows for easy identification. Every sandstone, siltstone and conglomerate in the valley is made up almost entirely of quartz particles, few if any show this crystalline characteristic. The quartz particles in the sandstone are only broken remnants of larger particles which have been tumbled, broken further, and rounded in their journey from one sandstone to the next. Pyrite, on the other hand, has formed since deposition of the rocks occurred but is usually found in small concentrations along bedding plains of coal seams or as a local replacement mineral in combination with clay or plant remains (small sticks and branches) in sandstone — the cubic shape in these instances can only be seen with the aid of a microscope. Calcite, likewise, shows up more commonly as a chalky amorphous (no particular shape) coating bonding sand grains together. Each of these minerals will, however, occur in their crystalline shape when precipitated from ground water slowly percolating through a rock cavity — a place where brand new crystals have space enough to grow (develop).

With the help of a 10 power, hand magnifying glass, second only to a rock hammer as the geologist's best friend, the individual quartz grains of a sandstone may easily be seen. Some sandstones will show grains more rounded than others, indicating the grains washed in from far away and spent a long time in the water before they finally came to rest, while others with highly-angular grains indicate a much closer source and less time in the water.

Occurring naturally in only rare instances in the Kanawha Valley, some of the many varieties of non-crystalline quartz (amorphous, hardened blobs of quartz) are chalcedony and agate (these make up the majority of the petrified wood in Arizona along with opal, another form of quartz) which form the banded and colorful geodes found cut and polished in rock and gem shops; be careful though because many of the cut "commercial" geodes have been artificially colored and do not reflect the geode's true subdued colors. Beware especially of geodes showing concentric bands of bright colors such as red, blue, and purple. For unknown reasons, store owners billing their wares as natural think customers would rather have an artificially (fake) colored and polished geode than a naturally colored one. I can barely look at these ancient wonders of another time; now sliced, dyed and commercialized.

Opal, another variety of amorphous quartz, is considered a prized gemstone. Onyx, a variety of chalcedony most often noted for its jet black color; jasper, that red gemstone sometimes called bloodstone; and Tiger's eye, the spectacularly striped and fibrous-appearing yellow and brown beauty are all varieties of amorphous quartz. Even crystalline quartz itself occurs in an array of different colors caused by impurities sucked from the water into its molecular structure when forming. Impurities such as iron, copper, and manganese, to name a few, which color the quartz crystals purple or bluish-violet, resulting in a variety known as amethyst, or the dark brown to gray variety called smokey quartz, and the yellow variety called citrine — and many more.

Of all the varieties of quartz on this planet, they all have one very simple thing in common, and that is their hardness. If you find a mineral in the Kanawha Valley and are completely stumped about what it is, check its hardness (the third identifying trait mentioned above), all you need do is see if it will scratch glass — if it does, it's quartz, if it does not, it is not quartz. Any sandstone in the valley will scratch glass because it is an accumulation of quartz particles, some silty shales will even scratch glass because they possess an abundance of fine quartz silt sized particles, and of course the Kanawha Black Flint discussed in Chapter Three will also easily scratch glass, because flint is non-crystalline quartz.

All minerals that occur naturally in the earth's crust have their own degree of hardness. A standard of mineral hardness, called the Mohs Scale of Hardness, was developed to rate and help identify the hundreds of different minerals ranging from the softest to the hardest; the softest in Nature being a one (1) and the hardest being a ten (10). What this means is each mineral rated higher (in number) then the previously one is harder, or will "scratch" the previous one.

Hardness	Mineral	Other
1	Talc	
2	Gypsum	Fingernail
3	Calcite	Copper penny
4	Fluorite	
5	Apatite	Glass, knife blade
6	Orthoclase	Pyrite
7	Quartz	
8	Topaz	
9	Corundum	
10	Diamond	Nothing

Table 2 — Mohs Scale of Hardness

In Table 2 Mohs Scale of Hardness is listed. Most of these minerals have nothing at all to do with the Kanawha Valley because they do not occur here but the table describes how several of the local common minerals relate to all the others.

For instance, listed at the top of the table is the softest naturally occurring mineral on the planet, which is talc (#1), and diamond (#10), listed at the bottom, is the hardest. All the other minerals the world over have been rated somewhere in between. Although the table doesn't list all the other minerals, it does list eight more to which others can be compared and identification can be aided. Notice from the table that I took some liberties and added several "other" objects relevant to our discussion and where they would fit if Mohs (whoever Mohs was) had decided to put them in. Note that a fingernail has a hardness of 2.5, and a penny is rated at a hardness of 3.

Talc is a very soft rock derived from the chemical weathering of basic igneous rock and used commercially in ceramics, dusting powders, plastics, etc; something to think about the next time you use your shower and bath powder. Diamond, at the bottom of the scale, is the hardest natural substance known. Diamonds are formed and found in volcanic pipes and are also found in placer deposits, which are just mineral concentrations in gravels and sands of river or stream channels, much like gold placer deposits — it's not the "mother lode" but the place where the minerals have gathered after being weathered from the "pipes" and washed downstream.

Diamonds have been found in glacial outwash gravel deposits in

WEEPING SANDSTONE

Michigan and Ohio and most likely many are still mixed up with the gravels of the Ohio River and its floodplain. It has been reported that in 1928, while playing horseshoes with his brothers at their home near Peterstown, Monroe County, William "Punch" Jones kicked up the third largest diamond ever found in the United States. It is suspected the diamond washed into the Jones' family yard from the eastern highlands of Virginia or North Carolina millions of years prior to the rather eventful horseshoe game. I'm not at all sure, but I suspect old Punch considered himself the winner. You just never know. It is also possible the diamond was carried there by Native Americans and left there for some reason.

Above I said all minerals have been tested and rated in accordance to their hardness, and this is true. All minerals are rated relative to Mohs Scale; for example, quartz just happens to be one of the privileged seven Mohs selected to be included in his scale of hardness. Pyrite, or fool's gold, is not listed in Mohs Scale but has a hardness somewhere between quartz and orthoclase (a white or pink mineral that makes granite colorful); so pyrite has a hardness of about 6.5 (see "Other" in the Table). Gold, on the other hand (not listed), has a hardness of only 2.5 to 3, quite soft and close to the same hardness as calcite (one of the ten minerals listed). In other words, most any mineral will scratch gold.

Common glass is important to mineral identification, at least in the Kanawha Valley, since quartz is the only mineral routinely found in the valley that will scratch glass (glass is listed as around 5.5 — quartz is listed as seven). Another test is to try scratching the mineral with a knife blade (actually a little softer than glass); the knife will not scratch quartz or pyrite but will scratch calcite. In many cases when a light-colored mineral is found in the valley, it is difficult to tell whether it is quartz or calcite, but if it can be scratched with a pocket knife most likely it is calcite. And, according to Mohs Scale, nothing will scratch a diamond (listed as a ten) but a diamond will scratch everything else.

Pyrite is probably the easiest of all to identify (with the noted exception of bright yellow sulfur) because it is always gold colored even if the crystals can not be seen clearly. Pyrite is commonly found in limestone, some dark organic shales, and especially around high sulphur coal seams. In the rocks of the Kanawha Valley, if a mineral

Plate 25 — A *Lepidodendron* cast Joe and I found some years ago in the old quarry across from the University of Charleston. It has since been destroyed.

is found that looks like gold and is associated with the other rocks of the valley it is pyrite.

Lycopods (Scale Trees)

During the Pennsylvanian Period, there were thousands of plant species, small ground-hugging plants, climbing plants that used the rocks and other plants for support, and tree sized plants. Lycopods, commonly called scale trees because their outside, or bark, surface looked something like the scales of a reptile, were prolific donors to the peat-gathering bogs and forest swamps. Two common scale trees, or at least their parts, found as fossils in the Kanawha Valley are *Lepidodendron* and *Sigillaria*. Lycopod leaves and cones are preserved as impressions; stems and roots are found as casts or molds, and more rarely the larger trunks are preserved as casts — some still standing upright in the same spot where they grew.

The trunk cast shown in Plate 25 is that of *Lepidodendron* found just above the Stockton-Lewiston Coal. The dark gray to black material surrounding the cast is carbonaceous shale, or what was once mud layers accumulating up from the base of the tree. The shale layers along the left side of the cast seem to "ride up" against the tree but are rather horizontal on the right side. This would be expected as the mud built up on the side that was obstructing the slow movement of flow which is a clue to the direction of the water movement — hence, left to right. Also, notice where the lower part of the outer surface of the cast has broken off that the sediment inside the cast is of the same type of dark organic shale but layered in a different orientation. This may indicate the tree trunk was rotted out and hollow and broken off a few feet above the level of the water, allowing the mud to spill over the top thus filling the cavity before the whole thing rotted away.

WEEPING SANDSTONE

We have all seen recently flooded reservoirs or reintroduced swamplands where dead trees are sticking out of the water, is this what happened to this tree? Where did the top of the tree go? It seems to just stop in the brown sandstone above. Whatever, this old tree grew and was buried right there on the spot where it was found. This is no illusion, this was once a real, live Lycopod tree now buried under hundreds of feet of rocks, and there are no more — the scale trees went extinct at the end of the Permian Period. When living, no squirrels scampered up the side of this tree and no birds nested in its high and safe branches. During the Pennsylvanian Period there were no such thing as squirrels and birds. The few animals that may have occupied its high branches were spiders and insects. About the only living relatives of the Lycopods today are the diminutive ground pines seen growing in starry clusters hugging moist hillsides in the company of ferns, and club moss.

Mature *Lepidodendrons* were very tall, lean trees towering to over a hundred feet with a ground level girth of only three to four feet (Figure 15). In spite of their majestic and tree-like look, the Lycopods possessed very little "wood." They were made up of relatively soft tissue organic matter — it's a wonder they lived as long as they did to get as big as they did and didn't blow over during a Pennsylvanian thunderstorm. Maybe because there were so many of them, along with all the other plant life of the time, that they were protected like the inner corn stocks in a cornfield. Also, because of the much higher carbon dioxide level in the atmosphere during the Pennsylvanian Period, plants actually grew faster; thus being able to replace themselves and grow

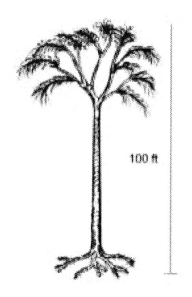

Figure 15 — A sky-high *Lepidodendron* tree of the Pennsylvanian Period. A shade tree it was not.

much quicker.

I mentioned above what one usually finds today as fossils of the lanky *Lepidodendron* (and *Sigillaria*) are its parts, which consist of its leaves, cones, stems, etc., and I also mentioned that these trees were called scale trees because of the appearance of their outside surface. Plate 26 shows the outside, or bark, surface of a *Lepidodendron* found in the Homewood Sandstone preserved as an impression in the medium to fine-grained sandstone. They really do resemble some type of scale — maybe scales from a giant snake. These scales, called leaf cushions, still remain pressed into the sandstone and represent the point of attachment of long needle-shaped leaves (several inches long on the trunk) that once grew spirally around the tree.

Plate 27 shows a close-up of the diamond-shaped leaf cushions of

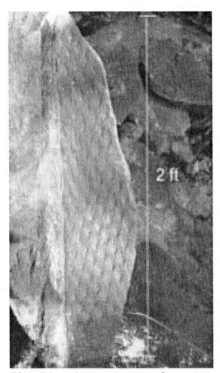

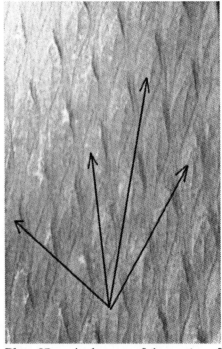

Plate 26 — A compression of a section of the trunk of the scale tree *Lepidodendron* preserved in the fine-grained Homewood sandstone.

Plate 27 — A close-up of the section of the *Lepidodendron* shown in Plate 26. Note the spiraling nature of the leaf cushions. There is no straight, up-and-down scale pattern.

the *Lepidodendron* shown in Plate 26. The spiraling nature of the leaf cushions is the identifying trademark of the *Lepidodendron* and distinguishes it from the other common scale tree, the *Sigillaria*. Some species of *Lepidodendron* have more "constricted" or flattened leaf scars, but always with the same spiraling configuration.

Lycopods, both *Lepidodendron* and *Sigillaria*, reproduced by spores released from cones growing from the tip of fertile branches. Plate 28 shows a compression of a *Lepidodendron* cone found in a thin (about one inch) layer of light-colored clay 15 feet below the Winifrede Coal, and a portion of a small stem with needle-like leaf shoots projecting from the leaf cushions. The cone appears almost like it has been cut in two exposing the cone's center stalk with the "hook-like" cone scales (bracts) extending away in both directions. The spores of the *Lepidodendron* were contained in bundles deep within these bracts. The grass or strap-like leaves of the *Lepidodendron* varied in length from an inch to several feet depending upon their location on the tree; the closer to the stem ends, the shorter they were. The leaves on the stem shown in Plate 28 are a little less than an inch long.

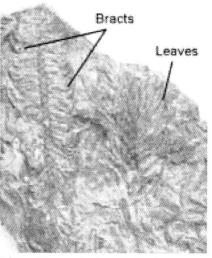

Plate 28 — A *Lepidodendron* cone (left) and a small branch preserved in a thin clay layer.

Lepidodendrons can be thought of as the oak trees of the Pennsylvanian Period. They were very successful and grew most anyplace where the soil and moisture was sufficient to sustain them. Growing fast and high, they towered over all other vegetation. If they lacked anything, however, it was "true" wood. A one-hundred-foot high *Lepidodendron* tree with a three-foot girth probably held enough wood the build a good chair frame. Today we have a tendency to think of a forest as being assigned to so many acres but there were no limits in the Pennsylvanian Period; there were no political, state, or county lines, no national forests. A forest stretched as far as it could,

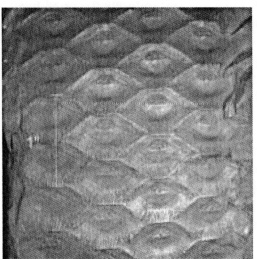

Plate 29 — A close-up of leaf cushions on a section of a *Sigillaria* trunk. Note the vertical arrangement of the leaf scars. Courtesy of the West Virginia State Museum Cultural Center.

only the environment stopped it.

Although not quite as commonly found as *Lepidodendron,* the other noteworthy and common Lycopod fossil found locally is the *Sigillaria*. *Sigillaria* was also a soft tissue tree and also had prominent leaf cushions but unlike the *Lepidodendrons*, which were arranged spirally around the stem, the leaf cushions of *Sigillaria* were arranged in vertical rows up and down the trunk and many species actually had vertical "ridges" separating the leaf cushions. When trying to differentiate between the Lycopods, these vertical ridges are most helpful; however, some species of *Sigillaria* had no lines and therefore must be studied carefully for the straight up and down pattern of the leaf cushions (Plate 29). The arrangement of the leaf scars also spiral like the *Lepidodendron* but are arranged vertically as well, something the *Lepidodendron* doesn't do. The compression shown in Plate 29 is in cannel coal which is a fine-grained, dense coal formed from the accumulation of trillions of spores and not the leafy debris so important to other coals.

Sigillaria was not much of a branching tree but instead grew clusters of longer lancelot (grass-like) leaves from common "bundles" which capped the tree. Cones full of spores hung from fertile branches waiting for the right breeze to shake them loose and carry them to the cluttered forest floor, though many instead drifted into the already thick, brown spore broth of the shallow swamp waters. Of the trillions and trillions of spores dropped from billions of Lycopod trees during the endless Pennsylvanian Period, what would science give for just one spore today? Although *Sigillaria* was a tall, slender tree it grew to only about 50 feet at maturity.

The tree-like Lycopods, so abundant during the Pennsylvanian Period, were major contributors to the coal seams mined today. They sucked up the sun's energy, stored it as carbon in their leaves and stems for 300 million years, and today they run our computers. Did all those trees grow and die so long ago just so we would have electricity today? After the coal is burned, the carbon, after being dormant for 300 million years, is once again (as mentioned before) liberated back into the atmosphere where it will in time once again be sucked up as carbon dioxide by another tree, bush, or the sea itself — the carbon cycle goes round and round.

Lycopods had a rather well-developed rootlike system which branched out almost horizontally in all directions from the trunk, each root branch having hundreds of water-absorbing rootlets projecting outward and arranged in a spiral pattern around the main root. Since roots are by nature already underground (no need for rapid burial), the roots of Lycopods are arguably the most common fossils found in the Kanawha Valley. In most cases Lycopod roots, called *Stigmaria*, occur as casts in siltstone and shales (ancient soils), and as compressions in some carbonaceous shales. Plate 30 shows a *Stigmaria* cast found in a stream channel just below the Winifrede Coal. More than likely the tree that was once hooked to the top of this root became a part of the accumulated plant debris of the Winifred Coal.

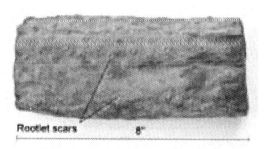

Plate 30 — A *Stigmaria* cast showing the spirally arranged scars where rootlets once were attached through which moisture entered the Lycopod tree. *Stigmaria* is considered the root of both the *Lepidodendron* and *Sigillaria* tree. Specimen about 8 inches long.

Although the Lycopods contained very little "real" wood, the cellular structure was nevertheless sufficiently durable to allow the trees to grow to amazing heights while growing mostly in the lowland swamp and delta areas.

Calamites

Another non-woody plant (tree) of the Pennsylvanian landscape and one that is a common fossil in the Kanawha Valley is the *Calamites*. Growing in thick stands along the margin of swamps and wide, meandering, shallow rivers, *Calamites* was also a first string player in the accumulation of peat-forming bogs of the Pennsylvanian Period.

Figure 16 is a representation of the *Calamites* tree, note that the limbs grew in whorls around a common "ring" on the trunk (called a node) and the leaves grew in whorls from around a common ring on the limbs (also a node) — a strange-looking tree indeed, but one of the easiest to identify when found as a fossil. Not all the nodes sprouted branches or leaves though, and like the Lycopods, the *Calamites* tree is found only in parts, that is, leaves are found, limbs and stems are found (Plate 31), and occasionally part of a trunk is found — many times still standing in place but incased in rock.

Figure 16 — *Calamites*, a strange-looking tree, is the extinct ancestor to the present day horsetails.

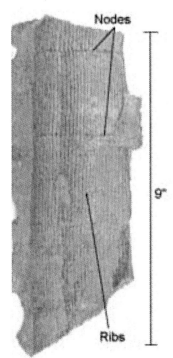

Plate 31 — A *Calamites* stem found just below the Winifrede Coal — Pottsville Series.

The stem cast in Plate 31 shows the ever present nodes and characteristic vertical lines (called ribs) that extend between the nodes. What you see in the picture is a siltstone cast of the internal pith region of *Calamites* and not the actual exterior bark features of the plant. After burial, the relatively large pith region within the stem decayed rapidly and filled with silt, thus preserving the tree's internal features — which are the most commonly found stem fossils of *Calamites* and easiest to identify.

The actual exterior or bark surface of the *Calamites* had less pronounced, or noticeable, ribs, and the nodes where the limbs were actually attached were made up of round branch scars that encircled the trunk or limb, as the case may be. Plate 31 clearly shows the detail of the nodal area (horizontal groves) and the ribs (vertical ridges) in between, and although some species of *Calamites* have the nodes closer together, some have them farther apart, and some show a greater distance between the ribs, all will hold this same characteristic pattern. The *Calamites* stem is probably the easiest of all the fossil stems to identify.

Calamites reproduced by spores which fell from fertile branch cones much like the Lycopods and had a root system (called rhizomes) much like cattails do today. The roots grew outward horizontally and at certain intervals sprouted a new *Calamites* tree. This method of growth along with new spore development led to dense and all encompassing *Calamites* forests along the water-saturated lagoonal margins — either marine or freshwater. I have seen artists' renditions of a typical Pennsylvania forest showing a random assortment of the common

Plate 32 — Leaf compressions of *Annularia* found in a thin clay layer below the Winifrede Coal, Pottsville age. The individual leaves are about one inch long.

trees of the time with a *Calamites* here and a *Calamites* there. Since I have never seen just one cattail growing amongst a variety of other plants, I can only wonder that there is something wrong with the picture. *Calamites* must have grown in great stands, or groves, completely dominating the surrounding vegetation.

The leaves of *Calamites* grew in successive whorls that encircled the stems and branches. They were only an inch or so long, and grew to the same length from each node. Like other trees of this period, when different parts were originally found as fossils, and nobody knew where all the parts went, they were given different names, such is the case with the *Calamites'* leaves. Although in reality there is only one brand of *Calamites'* leaf, there are two ways they can be found fossilized — and each way has a different name.

Plate 32 shows the leaves found in a "flower-like" or spread-out configuration. In this case, the leaves have been flattened perpendicular to the stem's axis; in other words, the leaves have been just flattened and spread out as if you are looking down on them from the top of the stem. When found in this whorled arrangement the leaves are called *Annularia,* and are commonly found associated with Lycopod parts, which implies of course that the Lycopods and *Calamites* tree types lived at the same time and probably competed for the best view of the swamp. Figure 17 is a representation of the whorled or spread-out leaf arrangement of *Annularia* as it existed in life and as it was preserved.

Plate 33 — *Calamites* leaves preserved in a "cupped" fashion around the stem are called *Asterophyllites.*

The second way, or fossil leaf configuration of *Calamites* leaves, called *Asterophyllites* (a big name for such a little leaf), is shown in Plate 33. In this case the leaves are sort of "cupped" up around the branch or stem, almost taking on a three dimension appearance. The leaves shown in Plate 33 are smaller than those shown in Plate 32 but represent the exact same thing only a different leaf arrangement when fossilized.

Figure 18 is a drawing of *Calamites* leaf whorls arranged in *Asterophyllites* fashion. Note that the leaves of *Calamites* are all the same (Figures 17 and 18) only the arrangement of their final resting place is different. The odd thing about all this is the rarity of both type leaf configurations being found as fossils on one stem, why is this? More than likely it has something to do with the orientation of the stem with respect to the direction and speed (if any) of the water current in which the stem finds itself. If the direction of water flow is from the base of the stem the leaves would have a tendency to wrap up around the stem and be buried that way; if the direction of the water flow is from the top of the stem, or little to no water flow at all, the tendency of the leaves would be to flatten out in a more natural position.

This hypothesis has no scientific basis at all, it just seems to make sense. To find these fossils, including the Lycopod leaves and cones, look in fine grained sediment like clays, shales and silty shales. Being rather fragile and containing no hard parts leaves are not preserved well in coarse grained sediment.

Lycopods and *Calamites* casts are rather easy to differentiate, Lycopods having mostly triangular shaped leaf scars of various sizes and *Calamites* almost always having the concentric nodes and vertical ribs — neither of which would have made very good backyard shade trees or wood for the fireplace.

In Chapter Three, while discussing the origin of enough silica to turn a shale into the Kanawha Black Flint, I mentioned the *Calamites* as a possible

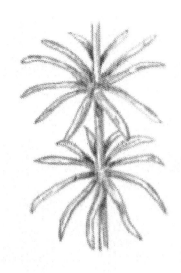

Figure 17 — Leaves of the *Calamites* tree arranged in death as *Annularia*.

Figure 18 — Leaves of the *Calamites* tree arranged in death as *Asterophyllites*.

source. Recall that the *Calamites* tree held within its tissue considerable quantities of minute grains of silica (today's scouring rush) that would be liberated to the water with the decomposition of the plant. Considering the vast forests of *Calamites* that surely existed and the shear bulk of the plant, some exceptional ones grew to 80 feet, it seems there would be sufficient introduced silica to go around and ultimately infiltrate the shale that would become the Kanawha Black Flint.

The tree sized Lycopods, which include *Lepidodendron and Sigillaria*, and *Calamites,* died out by the end of the Permian Period. The swamps were gone, the cyclic inundation of the sea had all but stopped, and the climate was slowly changing from subtropical to temperate (a more moderate climate). The surviving Lycopods (after the Permian) were runts compared to the great swamp trees, but persisted anyway throughout the millions of years of the Mesozoic and Cenozoic and continue today as ground pine and club moss. The same is true of the *Calamites* which exists today as modest horsetails and rushes at most three feet high and found along the banks of soggy creek bottoms and moist lowland meadows. These are truly living fossils. Their direct family ancestry exceeds most living organisms today. Although smaller and less imposing, these living fossils cover the damp ground in delicate, lace-like greenery (especially the ground pines).

The Lycopod and *Calamites* trees are an important part of our past. The black coal seams so conspicuous in the road cuts are the repository of millions of them, and although they seemingly offer nothing more to us today than just being incidental ground creepers and stubborn little survivors, they are still an important living member of the ecosystem that many of us seem not to notice.

Cordaites (The "Wood" Tree)

In the late Devonian and into the early Mississippian Period, as mentioned in Chapter Three, the first small, true "woody" plants began to appear in the fossil record. Unlike the soft-tissue Lycopods which reproduced by spores, these small trees had developed a vascular system consisting of, among other modern day tree-like features, secondary xylem — a botanists' term for woody tissue — and reproduced by seeds. By the time of the Pennsylvanian Period, these now extinct "hard" wood trees were growing to great heights, some well over a hundred feet and three to four feet in diameter, in widespread and luxuriant forests in the highlands above the swamps and river margins.

A highland, in this case existing in the vast Appalachian Geosyncline, may have been only a few tens of feet above the local swamp level. It has been suggested, however, that at least one species grew mangrove style in the shallow swamps and estuaries. These majestic plants as a group, the probable great granddaddy of the present day conifers (pines, hemlocks, redwoods, sequoias, etc.), were the *Cordaites.* There have been several names given to describe the Paleozoic Era transitional conifer wood itself, and it's difficult to find some kind of consensus among paleobotanists, especially since each part of the tree is given a different name. For this reason I will use the most frequently used name — *Cordaites.*

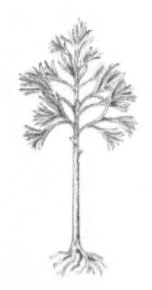

Figure 19 — A representation of the *Cordaites* tree. The *Cordaites* possessed wood and reproduced by seeds borne in cones.

At first glance (Figure 19) the *Cordaites* seems an unlikely choice to be the ancestor of the lofty pine, spruce and all the rest of the conifers. *Cordaites* had large, strap-like leaves up to three feet long and four or five inches wide which resembled corn stock leaves, while modern day conifers have small needle-like leaves at most eight to nine inches long. Also, *Cordaites* had an internal pith region within its young stems. By contrast, conifers do not have a central pith cavity. But the similarities outnumber the dissimilarities, and many of the similarities can be seen only with the aid of a microscope deep within the cellular fabric of the wood itself.

Cordaites reproduced by small, heart-shaped, winged seeds (Plate

34) borne from cones which grew off the main stems and branches, called *Cardiocarpus*. These seeds represent one of the earliest known seed plants and their ancient ancestors may have been the ancestors of even the seed ferns, so prominent themselves in the Pennsylvanian Period to be discussed later.

Plate 34 — A fossil "winged" seed from the *Cordaites*. Seed about one-half inch high.

Young branches and stems of *Cordaites* had a pith center consisting of a series of thin, horizontal membranes, or diaphragms if you will, which upon the death of the plant decayed quite rapidly leaving the hollow core as a mold to fill with fine sediment which, over time, formed a cast. The cast of the pith core when found as a fossil is called *Artisia*. *Artisia* has horizontal ridges throughout the vertical extent of the cast — these are remnants of the soft tissue diaphragms that bridged the expanse of the pith cavity. On occasion, *Artisia* is found with inclined ridges suggesting rapid stem growth where the diaphragms of the pith cavity could not keep up, thus stretching them in a ramp-like configuration.

Recall from Chapter Three while discussing the basal member of the Monongahela Series, specifically the Mahoning Sandstone, I mentioned that petrified wood is preserved in this member eight to ten feet above its base, and indeed it is. I also mentioned that these petrified trees are 90 million years older than the trees exposed in the Petrified Forest of Arizona, and indeed they are. The petrified tree logs found squeezed within the lower (early) part of the Mahoning Sandstone are the *Cordaites*.

I have already discussed the differences between stem casts and real petrified wood — casts having no preserved internal plant structure while petrified wood shows much of the internal structure as it was in life. Many of these old logs have been exposed in recent road cuts, some are more easy to see because they are almost black — suggesting the trees may have burned before being silicified. In other areas where exposed still in place, or when found in stream rubble, the wood is of various shades of brown to almost white depending upon the level of wood (lignin) still locked in the specimen and/or the amount of iron oxides present.

The predominant chemical components of woody tissue are cellulose and lignin, the former making up the basic structure of the wood — giving wood its elastic nature — and the latter, lignin, which is entwined around and within the cellulose, giving wood its strength. Cellulose decays more rapidly than lignin and given the right conditions is thus the first to decompose, especially when waterlogged or in a much depleted oxygen environment, which is the presumed final fate of the petrified logs of the Mahoning Sandstone. By the time the silica-rich (colloidal silica) water began to encapsulate the wood tissue, much, but not necessarily all, of the cellulose was gone, leaving only the cellular framework retained by the lignin.

There is only one way for trees to get locked up in such a hard place as we find these trees in place today, and that is if the trees were put there when the sandstone was sand. When these trees were felled, for whatever reason, sand was being washed in from the eastern mountains in great quantities and deposited in a sweeping delta extending far into the Appalachian Geosyncline (when the Mahoning Sandstone was being deposited). Something catastrophic must have happened in the proximity of a large stand of *Cordaites* trees to cause such a number to have been toppled and subsequently buried in the loose delta sand and all at about the same horizon. What happened? There are only so many ways a great stand of trees can be toppled at the same time; maybe a great flood washed down from the mountains ripping out the trees and carrying them to the point of burial. It's possible a nearby volcano exploded, knocking them over and into a handy body of water, but there appears to be no geologic evidence of volcanic activity associated with the logs. Maybe a meteor slammed into the earth in the vicinity of the trees but, again, there appears to be no geologic evidence of this either. A hurricane maybe, or an earthquake?

Because there are several inches of conglomerate found a foot or so below the wood horizon, which seems to be consistent in most sites where the wood has been found exposed, suggests a more rapid movement of water for only a short time before burial of the trees — this would lend some credence, but not much, to a destructive flood theory but would not explain the "burnt" wood. Whatever happened, large numbers of the *Cordaites* were uprooted and swept along with a swift current, all the while splintering and debarking from collisions with themselves and bottom rock, eventually piling up in some prehistoric and widespread logjam. One by one, as the days and weeks passed, the fragmented trees became waterlogged and in turn plunged onto the sandy river bottom only to be overwhelmed by more incoming sand. These once-proud trees, now fragmented and stripped of their bark and limbs, settled onto the turbid river bottom and reluctantly gave up the ghost.

The tree fragments and scarred trunks and dismembered limbs, however, did not become just another collection of casts as so many trees had done before. The water seeping down through the sand and into the already water-soaked fragments contained a relatively high concentration of colloidal silica (extremely small particles). This silica-rich water penetrated and infiltrated throughout even the deepest and smallest recesses of the wood, reacting chemically with the decaying wood (probably with whatever cellulose remained) and causing the silica to precipitate out of solution and begin to "frost" over even the most minute and microscopic cell detail. As the silica covered the wood fibers and filled up the empty cell columns with solid quartz, (chalcedony or agate) the decay of the plant slowed to nothing — the tree was turning to rock. Once completely "petrified," the tree or fragment thereof became just as indestructible as the quartz sand particles surrounding it, and in the process preserved for us images of unbelievable beauty.

Plate 35 — A typical Kanawha Valley section of petrified wood found in a small, rubble-strewn stream bed.

Plate 35 shows a fragment broken from a much larger slab of

petrified wood found in a stream channel that had cut through the Mahoning Sandstone. This particular relic is light brown, indicating little to no iron oxides present but still containing much of the original 300-million-year-old wood fibers (as mentioned above) still captured in quartz. Remove the quartz from this specimen leaving only the wood fibers and it would still burn.

Residents of the Kanawha Valley have been finding chunks of these trees in the tributaries of Kanawha River for years, especially in lower Davis Creek, Trace Fork, Rays Branch, Long Branch, Middle Fork, Little Creek, and all the small unnamed tributaries thereto. In fact, so much has been found that many residents have made stone walls out of the wood, or used large chunks as lawn ornaments and conversation pieces for rock gardens. It seems anywhere in the Charleston/South Charleston area where the lower Mahoning Sandstone is above drainage, the petrified wood can be found in the channel of the cutting stream.

A great flattened forest of *Cordaites* petrified wood has been weathering back out of the Mahoning Sandstone and into the creeks of the Kanawha Valley since the last regional uplift during the latter part of the Tertiary Period some four or five million years ago.

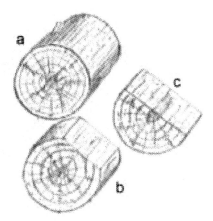

Figure 20 — Three different sections cut from log. a. Transverse (or cross section), b. Tangential, c. Radial.

Why is a tree that went extinct so long ago (in the Permian) thought to be the probable prehistoric ancestor to all the modern-day conifers? In spite of its large leaves, *Cordaites*, as already mentioned, reproduced by seeds borne in cones from branches much like conifers do today. Also, and much less obvious, is its internal cellular arrangement. I mentioned earlier that *Cordaites* was one of the first plants to possess sufficient quantities of "woody" (secondary xylem) tissue which allow its comparison to other plants. Different plant families have different cellular design. In other words, the cell structure of an apple tree doesn't look like that of a palm tree, and that of a palm tree doesn't look like that of a scotch pine. Once the cell structure is known, then the tree, or plant, may be identified as a group by looking at the cell structure under a microscope, and since a plant's cell structure is in three dimensions, a stem or branch must be cut on three different angles to get a good look — and this goes for petrified wood as well (Figure 20).

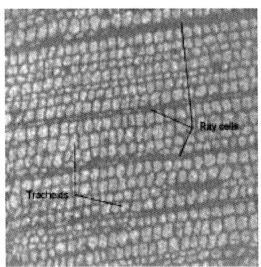

Plate 36 — A transverse section of the silicified wood of *Cordaites* magnified about 200 times.

First, the specimen must be cut *across* the stem or branch which results in a transverse or cross section cut (Figure 20a). In affect, the severed ends of the elongated vertical cells that run up and down the plant (called tracheids) — used for support and water transport — can be viewed. An example of a transverse section cut from a piece of the *Cordaites* petrified wood found in the Kanawha Valley is seen in Plate 36. What you see in Plate 36 is not new wood, but instead 300-million-year-old solid agate (quartz) "wood" that will scratch a hunk of glass as easily as a clear quartz crystal (remember, quartz has a hardness of seven). The cell structure is meticulously preserved in every detail; these very same tracheids, before they were filled with quartz, carried life-sustaining fluid and nutrients from a Pennsylvanian-age soil up to the lofty treetop of some ancient ancestor of the sequoia. There is no petrified wood on the face of the earth better preserved than that shown in Plate 36, and it's lying all over the place in the Kanawha Valley.

The cell structure as shown is a typical layout for what is called secondary xylem (wood), the round-shaped tracheids (like looking down on rows of tiny soda straws), now filled with hard quartz, are lined up in single file rows running from the plant's center, say to the right side of the picture, to the outside (bark side) to the left side of the picture. The darker "lines" running right to left and found in between every four to seven rows of tracheids are the horizontal ray cells, depending upon whether these shorter horizontal ray cells are one cell wide (like in Plate 36) or multiple will aid in distinguishing between tree species.

All the dark areas surrounding the individual tracheids are the actual cell walls which still contain wood elements but are now wrapped in harder-than-steel quartz. Each time I look at a newly prepared section of this wood I can hardly believe my eyes. Not only am I seeing an incredible work of Nature, but I am witness to what once was a living, juice-flowing organism that may as well now be from another planet, what's the difference between 300 million years and 300 million miles?

The other two cut types shown in Figure 20 allow different perspective views of the cell and ray walls when viewed under a high-powered microscope. Suffice to say, given the scope of this book, that the overall cell structure is very similar to modern-day conifer wood.

The macro and microscopic characteristics of the petrified wood found in the Mahoning Sandstone and the many tributaries to the Kanawha River indeed suggest the existence 300 million years ago of a transitional conifer. Like a good mystery story all clues point to a vast *Cordaites* forest (referring to the whole tree and not just the wood) leveled for whatever reason, captured in sand, and sentenced to an eternity locked up in quartz.

The microscopic picture seen in Plate 36 was made of only a small piece of a cut section of the petrified wood. This small piece was mounted onto a glass slide with special glue and was then ground down and polished by hand so thin that light would shine through it (called a thin section), enabling the cell structure to be viewed through the microscope.

The trees we have discussed thus far were significant contributors to the coal seams so noticeable today in road cuts from Charleston to Danville along Route 119, or up Greenbrier Street toward the airport, or wherever. There were several different species of both *Lepidodendron* and *Sigillaria,* each with slightly different leaf scar configurations but all with the same basic spiraling and straight up-and-down pattern respectively. The same is true with the *Calamites,* as mentioned above. Although *Cordaites* was a tree sized woody plant, there were also smaller bush and shrub plants that held woody tissue, all of which contributed to rapid accumulation of decaying swamp vegetation. Other tree sized Lycopods (besides *Lepidodendron* and *Sigillaria*) also existed during the Pennsylvanian Period but their fossils are less frequently found in the Kanawha Valley.

What words can describe the miraculous beauty of these old trees, crowned by the petrified wood of *Cordaites?* They grew when Nature let them grow and died when Nature let them down. The lineage of *Cordaites*, however, is still hidden in the DNA of the tallest trees on earth — the red woods of California, and in every common Christmas tree.

Chapter 5
Ferns, Vines, and Fern-Like Fossils

For the scent of a trail glazed over with frost,
and ten million times ten that have been lost.
All things must happen from A to Z....
it's Nature's way, can't you see!

Vegetation growth was totally out of control during the 40 million years of the Pennsylvanian Period. Atmospheric carbon dioxide was more than plentiful — much higher than it is today — enabling plant life to explode and spread to every available niche. Under a canopy of tall trees grew thickets of shrubs and blankets of creeping and climbing vines straining for every glitter of free sunshine; rotting plant litter gathered on the wet forest floor and oozed into the already debris-covered swamps and marsh waters, turning them black with organic muck. Spiders and insects, and flat, tank-like cockroaches dined on the putrefying vegetation and each other, and grew to enormous size (for bugs). The weather was hot, sticky, and humid, and there were no paths or wooden walkways. This is the way it was, not in some far off corner of a South American jungle, but right here in the Kanawha Valley — and the proof is all around us.

Ferns

Second only to the Lycopods as the most prolific foliage on the Pennsylvanian landscape were the ferns, some of which grew to a height of 90 feet while others grew as understory plants filling whatever space was left. Akin to the small, arching ferns seen growing over the shady, moist forest floor today, Pennsylvanian-age ferns reproduced by spores which grew in spore sacks on the underside of fertile leaves. Large, flowerless fronds spilled out in all directions from the top of the tree fern much like palm and coconut trees do today; except, unlike the palms and coconut trees, each tree fern frond divided and subdivided into multiple smaller leaves giving the frond a delicate lace-like appearance.

Although tree ferns resemble palm trees, they are from totally different plant groups. Palms have flowers and reproduce by seeds (some big seeds like the coconut) while ferns reproduce by tiny spores. Also, many palm trees have long, strap-like leaves while the fern's fronds possess multiple leaves and many tiny leaflets (Plate 37).

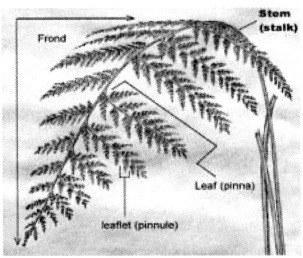

Plate 37 — A cloudy day in fernville. A representation of a single fern frond showing the names of the different parts. This nomenclature is used for ferns in general, whether tree size or ground huggers.

Paleobotanists, scientists who study ancient plants, group all Pennsylvanian-age tree ferns under one form of growth morphology (tree design) called *Psaronius*. In other words, there are many different kinds of ancient tree ferns, all of which have different leaf or frond designs which will be discussed in a minute, but only one basic type of overall tree design (called *Psaronius*). From a distance they must have all looked alike — as a matter of fact they looked exactly like tree ferns do today. *Psaronius* looked much like the tree ferns found in the modern-day tropics, and they should because *Psaronius* is the ancestor of the modern-day tree ferns, although tree ferns today reach a maximum height of only about 65 feet (90 feet for *Psaronius*). Imagine a 90-foot high palm tree capped with fern foliage, roughly the height of a nine story building! Not only were there tree sized ferns during the eternity of the Pennsylvanian Period, there were herbaceous (bush sized) ferns, ground ferns, and climbing ferns, all of which contributed vast amounts of dark organic matter to the swamps, lakes, lagoons, and any other nearby harbored body of water.

Psaronius is a catch-all name for all the tree ferns of the Pennsylvanian and Permian Periods, and all the various kinds of leaves found on *Psaronius* are placed in the genus *Pecopteris.* The actual differences found in the leaves, or leaflets, give rise to the various species of *Pecopteris* which are in turn, and for the most part, determined by the shape of the leaves, or leaflets, the vein patterns, and, in a few instances, branch and stem features. In other words, all the true tree ferns had the same first name: *Pecopteris,* which is the "genus" name. Based on the differences I just mentioned, the fern's second name, which is the "species" name, can be determined.

Plate 38 shows a typical *Pecopteris* leaf with multiple small leaflets attached to the stem by their entire base (a *Pecopteris* trait) and the sides of the leaflets almost parallel to one another. Also, the leaflets themselves are almost always the same length and, when viewed from under the microscope, the leaflet's vein pattern shows a prominent midvein running all the way from the bottom (where it attaches to the stem) to the top of the leaflet with lateral veins branching off the midvein at about 45 degrees and forking into two veins prior to reaching the leaflet's outside edge. Once all the leaflet features have been considered the species can be identified, and in this case (Plate 38) the fern is identified as a *Pecopteris hemitelioides* (I think) which grew on the *Psaronius* tree. It is difficult at best in most cases to identify the species of a fossil plant, and unless you are a college professor in paleobotany, totally unnecessary.

Plate 38 — Single *Pecopteris* leaf with multiple leaflets.

There are many different species of *Pecopteris* based on the plant features I mentioned above, and the reason I went through this exercise of species identification was to familiarize the reader with the process one goes through and what to look for when trying to identify a particular plant fossil.

Although not a museum specimen by any means, the leaf portion shown in Plate 39 is a compression of another fern which resembles *Pecopteris* but most likely represents the underside of a leaf, showing spore sacks, called *Danaeides,* which supposedly was a herbaceous (bushy) fern. In this case, the base of the leaflets attach to the stem by a wide single point at the midvein, the leaflet's sides are nearly parallel, the tip end is rounded or blunt shaped, and the lateral veins extending from the midvein to the leaf edge are nearly perpendicular to the midvein.

Plate 39 — A herbaceous fern leaf, collected from the Pittsburgh Red Shales.

Maybe not a real good specimen of an ancient fern by any means, but the interesting thing (at least to me) about the fern shown in Plate 39 and the real reason I wanted to include it at all is that it was collected from the Pittsburgh Red Shales several hundred feet up into the Conemaugh Series. Red shales are notoriously unfossiliferous simply because of their mode of deposition (river flow deposition with frequent drying). Although I have found many unexpected plant fossils in these particular red beds in quite a small area, both part and counterpart, the black carbon usually associated with plant compressions found in most shales seems to be absent — obviously a result of the depositional environment and/or the chemical complexities between iron (hematite which turned the shales red) and carbon — and without the black carbon the plant features are much more difficult to see. It's also possible the carbon just leached out over the eons.

In any case, those residents living in the outcrop area of these red shales should start looking; just based on my own experience it is difficult to say just what may be found. The fossils of the red beds will consist of terrestrial plants and maybe other freshwater organisms such as gastropods (snails), although I have found the tiny, snail-like calcareous tubes of the marine worm *Spirorbis* in the Pittsburgh Red Shales (Plate 40).

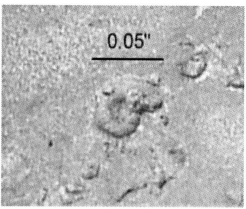

Plate 40 — A tiny *Spirorbis* (0.05" at its largest dimension) attached to an impression of a leafy plant.

These little marine plant eaters and their secreted tubes (it would take about twenty of them lined up in a row to make an inch) have been common from the Ordovician Period to the present, and in almost all cases are associated with marine sediments and other marine fossils, and are known to exist only in a marine environment today. They are found as fossils attached to seaweed compressions, brachiopods casts, marine gastropods, or anything else organic.

So, if *Spirorbis* is a marine worm, why is it found in the freshwater Pittsburgh Red Shales? This leads to four possible answers; first, maybe some species of the ancient *Spirorbis* adapted to freshwater living, or, second, on occasion the sea levels may have risen and encroached upon the red beds allowing *Spirorbis* to do its thing; third, this area of the Pittsburgh Red Shales was not freshwater deposits at all but formed in a marine environment; and fourth, the red beds were formed in a tidal or storm zone which allowed *Spirorbis* to occasionally come ashore and live in tidal or storm pools.

The first option is not good because the particular species shown in Plate 40 is found in marine sediments — the same species can't live in both sea water and freshwater. The second option isn't real good either because red beds, when under the sea very long, most likely will turn gray. And since I have found no other marine creatures in these shales (as yet) and it is full of terrestrial plants, the third option is remote as well. At this point in time I choose the forth option — this choice may change as I continue to study and collect from these curious red beds. It takes a good eye to see one of these miniature wonders and then a good hand-lens to confirm it.

Fern-Like Seed Plants

Keep in mind that there were ferns (or true ferns because they reproduced by spores like today's ferns) during the long years of the Pennsylvanian Period and there were plants that reproduced by seeds and looked just like ferns, commonly called fern-like seed plants or just seed ferns.

During the Pennsylvanian Period there were only a few plant groups that reproduced by seeds, the *Cordaites* for one and a group of plants that looked for all the world like ferns, but were not even related. These plants have come to be known as seed ferns or plants with fern-like foliage, and although greatly outnumbered by the spore-bearing plants of the day, they left a remarkable assemblage of fossils behind and contributed heavily to the widespread swamps and future coal beds. As with the "true" ferns (just covered above), seed ferns came in all sizes from tree size to small shrubs.

The tree sized seed ferns, known collectively as *Medullosa* which were on average much smaller than the true ferns (*Psaronius*), grew to an average height of fifteen to twenty feet (it has been reported that a few outliers grew to a height of sixty feet) and produced an array of distinctly shaped fronds and leaves. Like the true ferns, seed ferns have been given different genus and species names and are identified by their leaf or leaflet shape, their vein pattern, how their leaves attach to the stem, and physical characteristics of the leaves themselves, such as leathery looking, thin, heavily veined, etc.

Although tree sized seed ferns, as well as true ferns, grew well along river, lake, and swamp margins, they grew equally well on the humid, misty slopes that leisurely drained into the lowland waterways. And with only Nature's do or die attitude to guide the growing frenzy, there must have been a brutal herbage battle for every inch of growing space, what with the multitude of big and little Lycopods, all sizes of *Calamites*, ferns and seed ferns of every size and shape, vast stands of the hard wood *Cordaites,* and a host of known and unknown understory plants crammed in for good measure — the mother of all impenetrable tropical forests. I suspect even Lewis Wetzel, West Virginia's celebrated frontiersman and trail-blazing scout, could not have penetrated these woods more than a few hundred yards simply because there was no place to start and no place to go.

It is difficult to imagine such a place stretching over thousands of square miles: plant life so dense and foreboding and all the while concealing down under the canopy fetid and slow-moving, scum-covered swamp waters; a sticky lacework of huge, thick spider webs spread out in every direction from stick to branch and whose lurking, eight-legged tenants were equally huge and nasty; an infestation of ugly bugs creeping, jumping, and flying in a swarming frenzy — eat, eat, eat. And the largest of all, the lumbering, flat-headed amphibian, menacing the shallow water margins gulping down with its wide slimy mouth anything that moved. Stout-bodied reptiles were there too — four legged, slow-moving creatures with thick skins prodding along the mud flats and reed-covered sand bars looking for washed-up dead fish or pushing their way through the steamy underbrush anticipating a entree of three-inch-long crunchy cockroaches.

What seems to us as an unbearably hot, humid, and hostile environment was in fact a booming wonderland of life; both the animals and the plants were experimenting with their own existence. Each born into a time of plenty with but one objective — reproduce before you die.

Today we take seeds for granted but it was not always that easy. The first land plants found in the fossil record of the Silurian Period were spore bearing, and this process of reproduction continued for 85 million years until the Mississippian Period where for the first time seeds begin to appear in the fossil record associated with *Cordaites* and primitive seed ferns. Since these seeds were "already seeds" who knows how long or in what form seed transition took from an even further ancient spore-bearing loner that by chance found it more advantageous to fertilize its spores while they remained high in the branches. Although spore-bearing plants did quite well during the unending years from the Silurian to even the present time, spore reproduction requires several more perilous steps than does seed reproduction.

First, a mature fern produces a cluster of tiny spore sacks called sporangium on the underside of its leaflets. When the time is right the spore sacks burst open spilling hundreds of microscopic spores onto the ground; if a particular spore is lucky it will fall or be blown into and under the existing ground vegetation to a shady, moist place where it will germinate into a tiny, green, leaf-like plant called a prothallus. Now, here is where it gets wild, this little heart-shaped prothallus, only about a quarter of an inch wide at maturity, develops female cells (eggs) on one edge of the plant surface and male cells (sperm) on the far edge. To the male cells, because of their microscopic size, the distance across the surface of the prothallus must look like the length of a football field. The male cells have to wait around (and they don't have that long to wait before both they and the prothallus wither and die) until there is sufficient moisture on the surface of the prothallus to float over to the female side for fertilization — at which time a new plant will germinate.

It is quite apparent that this reproductive strategy works well in a moist environment which was far reaching during the Pennsylvanian Period but not so commonplace during the previous Mississippian, and does not work much at all in a drier "upland" environment. Even today, ferns only grow on the shady, moist hillsides or deep in the shadowy valleys heavy with morning dew and around soggy creeks or lowland springs. Seeds, on the other hand, need only fall on damp ground, since they have already been fertilized up in the tree, set their fast-growing rootlet to get what moisture there is in the soil, and start growing. A seed can even fall on dry ground and wait for some time for a rain shower. This twist of reproductive adaptation freed plants to grow in places never before available to the spore population, a real evolutionary attraction during the swamp-less years of the late Devonian and early Mississippian Periods when they must have evolved.

Seed ferns may have evolved from an ancient pre-*Cordaites*-like plant of the Mississippian Period or they (*Cordaites* and seed ferns) both may have evolved from a common transitional seed bearing

WEEPING SANDSTONE

ancestor. In any case, fossil seed-like structures (Plate 20, Page 131) confused paleobotanists for a long time until they were found associated with *Medullosa* leaf fossils. Plate 12, back on Page 95 is a beautiful example of a compression of a portion of a leaf of the seed fern *Mariopteris*. Had this specimen, which is about four inches long, been green, one would think it had just fallen from the tree. This particular genus (*Mariopteris*) is noted for its rather prominent midvein, alternate and somewhat pointed tip leaflets, broadly attached to the stem by the midvein and part of the lower portion of the leaflet, and some lobing on the larger leaflets. The lateral veins of the leaflets branch away from the midvein at about a forty-five degree angle and then gently arch toward the edge with some of the veins forking along the way.

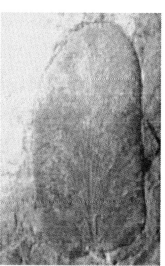

Plate 41 — *Neuropteris pocahontas* showing the midvein attachment to the stem just below the leaflet lobes. Found in silty shale above Coalburg coal, Pottsville Series.

Plate 42 — A *Neuropteris scheuchzeri* leaflet showing the shortened midvein and the sweeping nature of the lateral veins. About one inch long.

Another example of a seed fern, this one belonging to the genus *Neuropteris*, is seen in Plate 41. The leaflet attachment to the stem is by a single point at the base of the midvein which is one of the identifiable characteristics of *Neuropteris*. Generally the midvein of

Neuropteris is less apparent than many seed ferns and may only extend halfway through the leaflet while the lateral veins branch off from the midvein in a sweeping, arching manner, sometimes forking several times before reaching the outer edge (Plate 42).

The single leaflet shown in Plate 42 was found on an exposed bedding plane in a rock pile of rather large fragments of dark shale dumped by the side of the road several miles up Paint Creek. In a slab about two feet by three feet were roughly fifty of these individual leaflet compressions randomly scattered about — all of which seemed undamaged.

What circumstance could cause the detachment of multiple healthy leaflets from the stem of an apparently healthy seed fern and scatter them unharmed into muddy water below? Granted, this mystery may not be on par with the mystery of the disappearing dinosaurs but nevertheless represents a day in the life of the extinct *Medullosa* seed fern *Neuropteris*.

Anyone who reasoned it was the Fall of the year and the leaflets just fell off naturally made a good guess but is probably wrong. When these trees grew, West Virginia was situated just north of the Equator on the super continent of Laurasia and enjoyed a year-round tropical environment (no tree rings are found in *Cordaites* petrified wood). Had a flood occurred, the leaflets would more than likely still be connected to the stem such as those seen in Plates 12 and 41. Perhaps a wind storm liberated many leaflets from the tree, which is a common occurrence today, and blew them randomly into the shallow, muddy, fringe water of a freshwater lake...unfortunately, causes are rarely fossilized.

Every inch of vertical rock exposed or unseen in the Kanawha Valley is an effect, or result, of some geological or atmospheric activity such as erosion, deposition, storms, drought, earthquakes, volcanic activity, inundation or withdrawal of the sea, etc. When looking up close at an exposed road-cut or outcrop along a forest trail, look at the vertical and sometimes horizontal changes that occur within the body of rock itself which are testimony of the different geological or climatic changes that occurred when the rock was being deposited, things like grain size change, color change, actual rock type change (from shale to coal), rate of change (gradual or rapid), and layer thickness which reflect the relative time span of the deposition. The past is shrouded in mist only to the uninformed.

Alethopteris (Plate 43) is another common genus of the seed fern *Medullosa* found in the Kanawha Valley, particularly in Pottsville and Allegheny-age shales associated with coal beds, and is represented by several different species. Some of the distinguishing characteristics of *Alethopteris,* i.e., those things that make an *Alethopteris* an *Alethopteris,* are the thick, leathery-looking leaflets, a very prominent midvein extending from the base of the leaflet to the tip, prominent lateral veins extending almost perpendicular from the midvein to the leaflet's edge, and attachment to the stem by the entire base of the leaf which extends from one leaflet to the next lower leaflet — called a confluent base. *Alethopteris* is one of the most recognizable of the seed ferns. In some species the leaflets are shorter, some fatter and some closer together, but all generally have the same form details.

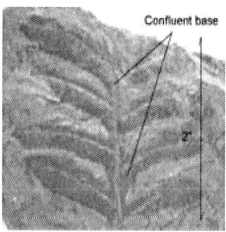

Plate 43 — *Alethopteris serlii* collected from a shale layer in the lower Allegheny Series.

Yet another seed fern which is represented by many species is the broad and varied genus *Sphenopteris* (Plate 44). There are so many species assigned to the genus *Sphenopteris* that if you find a plant fossil and have no idea what to call it, call it a *Sphenopteris* and you have a good chance of being correct. Some paleobotanists have divided the genus *Sphenopteris* into several genera, but for our purpose they will be left as *Sphenopteris* with the other genus in parenthesis where applicable.

Plate 44 shows a *Sphenopteris* (*Rhodea*) type foliage a bit unusual in that it has thin, flat leaflets branching several times from the primary stem. There are several of these thin-leaf type plants represented in the fossil record; some, in fact, are described as true ferns with only the smallest of details separating them, and the rest almost impossible to

Plate 44 — A small *Sphenopteris* (*Rhodea*) leaf of the Pottsville Series. *Rhodea* has thin, grass-like leaves unlike other *Sphenopteris*. 3/4 inch long.

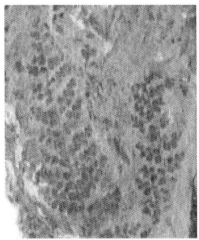

Plate 45 — Two *Sphenopteris* leaves taken from a thick, silty shale layer just below the Coalburg Coal, Pottsville age. Longest leaf about 5 inches long.

tell apart in the field.

The *Sphenopteris*(*Eusphenopteris*) shown in Plate 45 has multiple alternate branching leaflets (pinnules) which then also have smaller alternate and somewhat rounded secondary leaflets that attach to the stem at the midvein point. In most cases, the midvein and lateral veins are less than obvious without the aid of a hand lens. The alternate arrangement of the leaves just means that the leaves sprout from the stem in a "stair step-like" pattern up the stem and don't grow from the stem directly across from one another.

I have shown more seed ferns than true ferns in the Plates because I've found that the seed ferns are more numerous in the shales and fine-grained siltstones of the Kanawha Valley, and there are more leaf form distinctions between the seed ferns than the true ferns; at least in terms of "in the field" identification. The type seed cast shown in Plate 20, with several variations, has been found associated with and attached to fern-like foliage of the *Medullosa*, especially the *Neuropteris* and *Alethopteris* seed ferns.

As successful as the seed ferns were during the Pennsylvanian and into the Permian Period, for whatever reason, during the latter half of the Permian, their numbers began to dwindle and by the end of the period the seed ferns were extinct — and there has not been a *Medullosa* tree fern since.

Beginning in the middle part of the Permian Period, say, 265 million years ago, weather patterns were beginning to shift. The climate was slowly getting cooler and dryer as the eastern seaboard was being wrenched, buckled, and shoved into the air as the super-continent of Gondwana slowing crunched its way into the super-continent of Laurasia. As mentioned earlier, this colossal event, which lasted roughly 100 million years, also raised the Appalachian trough from an area of deposition to one of erosion — the Appalachian Plateau. Obviously, during the uplift it didn't take long for the swamps to begin to drain off and loose their store to erosion along with other recently deposited sediments.

In this slowly-changing environment one would think the seed ferns would have had an even greater chance of survival than the true ferns simply because of their mode of reproduction. But apparently it didn't matter what method of reproduction a plant had, since the Lycopods and *Calamites*, both spore-bearing plants, all but became extinct with just a few punitive species surviving as previously mentioned. Even the apparent granddaddy of all seed-bearing plants, the ancient and stately *Cordaites*, after diverging into several other seed-bearing plant species that did manage to survive (thank goodness), bit the dust.

Looking back on this time in the fossil record and talking about it makes it seem to have happened rather quickly, but it didn't. From one generation of plants *and* animals to the next, life went on pretty much as it always had. Extinction prowls around and stalks without the species ever realizing it, ever so slowly reducing their numbers from one range to the next. It comes not just when the last lonely tree keels over, it would die in due time anyway like all trees before it; extinction is a process that has a beginning and an ending — once seriously started there is no turning back. Extinction doesn't mean the day the last plant died; extinction means the day the last plant germinated.

During the slow climate change occurring in the Permian, the probability of germination and growth to maturity of many plant species lessened. New trees become fewer, with fewer trees comes fewer seeds and fewer seeds means even less chances for germination. Extinction eats on itself until there is nothing left.

Before leaving plants altogether, let's come down from the trees a moment and discuss a commonly found understory plant that spent its life in the shadow of the trees. This particular plant, a spore-bearing plant, is thought to be a scrambler, that is, it grew on long sinuous climbing vines with spiraling leaves at equal distances up the stem (sort of like the leaf whorls of *Calamites*, to which it was related) and sometimes used other plants, dead or alive, for support, although some species were bush size and seemed to be able to support themselves. These relatively small vine plants had a big name, *Sphenophyllum* (Plate 46), and are found as fossils today in many of the freshwater shales along with the leaves of all the other vegetation discussed so far.

A fact that may seem obvious yet should be mentioned anyway is when a plant (or animal) is found associated with another plant (or animal) these organisms lived at the same time and most likely in the same neighborhood. Many times when a fossil is found there will be other fossils found along the same bedding plane or at least within close proximity vertically. Gathering and identifying all the fossils within a particular layer, or zone, can lead to an understanding of the assemblage of life that lived together at the time of sediment deposition — or at least those who died together.

Plate 46 — *Sphenophyllum*, a small spore-bearing plant, most likely covered the forest floor and scrambled over dead logs and up the sides of larger vegetation. Common in Pottsville and Allegheny-age shales. About a inch across.

Like ferns and fern-like foliage, the identification of the various species of *Sphenophyllum* are based on leaf characteristics since, in most cases, the stems and branches all look alike. The species pictured in Plate 46 has leaf whorls made up of six leaves (can be up to eight), each leaf is fan shaped with tiny serrations (sawtooth-like) on the larger convex end, and veins fanning out from the stem end (sometimes forking a couple of times) to the tip. This particular plant has the full name of *Sphenophyllum emarginatum*.

For most of us, knowing the genus name for fossils is good enough — the different species names may not be that important. In many cases I believe the minute distinction between species exists only in the mind of the original describer. More than likely on an average size sugar maple tree there are no two leaves out of thousands that look exactly alike (makes sense to me), each leaf having subtle shape and vein differences from that of its neighbor. I am convinced that if all the leaves of a single sugar maple tree were scattered over a present day mudflat and fossilized, scientists of the future would discover one genus but a hundred species.

Only a few of the many locally available plant fossils have been discussed; for the reader who wants to know more about the identification of fossil plants of the Pennsylvanian Period I have listed several excellent books in the reference section. An interest in fossils is no different than an interest in anything else, you just can't explain it. But it's more than just finding a fossil (or buying one for that matter) and then putting it on a conspicuous shelf in your living room to be used as a conversation piece when guests come over; it's the total experience of knowing where you were geologically when you found it, what you have found, the fossil's place in prehistory, and most of all, why it was where you found it in the first place — and even how it got there.

Every single fossil is of scientific interest even though a thousand of the same type may have been found before. A *Calamites* cast found in, say the lower part of the Homewood Sandstone, looks a lot like one found in the upper part of East Lynn Sandstone but is vertically separated by 150 to 200 feet or more of sandstones, various shale layers, and at least one relatively thick coal seam (No. 5 Block). The two *Calamites* casts are not only separated by vertical distance but by time; several hundred thousand years may separate the death of these two *Calamites* and each one died and was subsequently buried in its own way — separated by countless years but not by lineage. The study and classification of plant fossils is called paleobotany and the science itself is the real joy of fossils.

BOB KESSLER

Fossil Animals

Animal fossils are quite rare in the Kanawha Valley when compared to plant fossils, but certainly not absent altogether. Plant fossils are, for the most part, found in freshwater deposits, and the majority of rocks found in the Kanawha Valley are of freshwater origin. Most of the sandstones and siltstones were deposited by rivers in one way or another, most of the shales were deposited in pre-swamp freshwater environments as were the clay and mudstone deposits. The coals, of course, were freshwater swamp deposits, and what few thin bedded limestones there are in the valley are mostly freshwater lake deposits. In general, all of these different freshwater rock types, excluding the limestones in the valley, contain scattered plant fossil remains all the way from nondescript stick fragments to leaf compressions to whole trunks still standing where they grew; lucky is the individual who finds a freshwater or terrestrial (land) animal fossil — but it can be done.

An example of a terrestrial animal being captured and fossilized in a freshwater shale deposit is shown in Plate 47. The fossil may be difficult to see at first but once you do the image is quite apparent.

This insect, possibly a small, ancient dragonfly or damselfly, or maybe even a wasp, became trapped in a silty mud flow along with a single leaf of *Sphenophyllum* (different species than that shown in Plate 46 because it has larger "teeth" on its big end). Perhaps the insect was resting on the leaf when suddenly submerged, or perhaps it was injured and couldn't escape even if it wanted to. Whatever occurred, the flying insect was bushwhacked in a freshwater mud bath some 300 million years ago and unwittingly left its carbon signature glued to its final resting place. This particular insect, only about a half inch long, had large, bulging eyes and two sets of delicately veined wings that apparently folded back onto the body when not in flight. Flying insect compressions are extremely rare because of their delicate nature and were more than likely eaten by other insects or animal scavengers before being buried.

A wild guess could be made that ninety-five percent of the rocks in the Kanawha Valley are of freshwater origin, yet freshwater animals like the freshwater gastropods (snails), bones or imprints of amphibians, early reptiles, and other insects are relatively rare. Its understandable that sandstone, because of its coarseness, does not preserve delicate fossils well but does hold many casts of tree stems and branches. Would not a bone be just as likely to be preserved as a limb from a soft tissue *Lepidodendron*?

During the Pennsylvanian Period there were a profusion of cockroaches, beetles, and insects — why aren't their remains in the freshwater shales more common? Presumably beetles, which include cockroaches and having rather hard chitinous exoskeletons, would be just as easily fossilized (if not easier) as the tender green leaves of the seed fern which are found in abundance in many shales. One could, I suppose, make the argument that in spite of the horrendous number of beetles making a living during the Pennsylvanian Period, there were a million times more leaves, thus the greater chance of a given leaf becoming fossilized — one percent of a ten trillion leaves is greater than one percent of a ten million beetles.

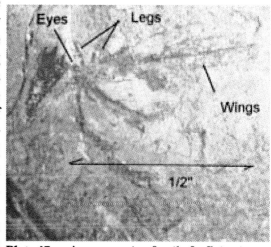

Plate 47 — A compression fossil of a flying insect found in a freshwater shale just below the Stockton-Lewiston Coal — Pottsville age.

Regardless of the apparent scarcity of freshwater animals in the valley (maybe I just haven't found many yet), there are several shales that contain a good assortment of marine fossils. Recall from Chapter One that the sea rose up over the Appalachian Geosyncline periodically and inundated the newest freshwater deposits. Sometimes only shallow seawater covered the land, sometimes it was deeper, sometimes the seawater was calm and in a protected cove, and sometimes it wasn't. The resultant marine deposits reflect the depositional environment as covered in Chapter Two. In most cases the marine deposits in the valley are dark organic shales (including the Kanawha Black Flint), silty shales, and mudstones which usually imply relatively calm, shallow water deposition. With the sea also came the sea animals.

Without a doubt, the most common marine animal fossil found in the valley is the brachiopod (Figure 5, Plate 21, Plate 23, and Plate 48), this hard-shelled, clam-like bi-valve of many genera and species was especially bountiful during the Pennsylvania age. Like the term trilobite, pelecypod, and gastropod, brachiopod is a common catchall name that describes the entire family of these remarkable creatures.

One, if not the most famous of the marine shales, which in some locations in the Kanawha Valley is six to seven feet thick, is the Kanawha Black Flint. This shale is not only an enigma because it now is a flint, but also a puzzle of sorts because it is not consistent in carrying fossils. In some exposures fossils seem to be plentiful, although difficult to get out, and in others nary a one can be found. Plate 48 shows a cast of a marine spiriferid brachiopod (another catchall name for deeply striated and wing-hinged brachiopods) found in the flint from an outcrop on Dry Branch. This is a beautiful specimen of an extinct spirifer that lived long before the Neanderthal learned the hard way not to throw their dead out in front of their caves but to bury them to keep the sabertooth cats or other big meat-eaters and scavengers away from their front door, and long before there was even a clue of T-Rex. Do you suppose the Neanderthals filled the "grave" up with flowers just to cover the smell?

Plate 48 — A spiriferid brachiopod collected from the Kanawha Black Flint, Dry Branch: about 3/4 inch high.

Another particularly abundant sea animal of the Pennsylvania Period was the pelecypod. Today pelecypods go by various names, such as, mussels, clams, bi-valves, scallops, and shellfish, depending on whether you are eating them or collecting them. Oysters, hunted for their meat and occasional booty, are also pelecypods. Living in great masses called oyster beds, they used to populate the undulating and marshy coastline of Chesapeake Bay and were a favorite staple of the bay area Native Americans. Pelecypods, like brachiopods, come in many shapes, sizes, and exterior design, and in spite of their great numbers in the Pennsylvanian seas they are not the most common fossil in the marine layers of the Kanawha Valley, yet I have found many different types.

Although there are external similarities between a pelecypod and the brachiopod, they are in fact not related and represent two completely different biological groups of sea animals. The pelecypod cast shown in Plate 49 is quite obviously not a brachiopod — why?

Notice that the white line drawn vertically on the shell is centered on the shell apex (at the top) and extends to the bottom of the shell, essentially cutting the shell into two parts. Now notice that the two sides are not symmetrical; that is, the left side is "smaller" than the right side. If you drew the same lines on the brachiopod in Plate 48, both sides would be the same, or, symmetrical — hence a sure fire way to tell the difference between a pelecypod and a brachiopod no matter what kind of markings are on the shell.

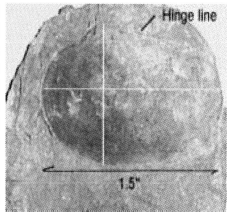

Plate 49 — An average size pelecypod found in marine shale some 40 feet below the Coalburg Coal — Pottsville Series: about 1.5 inches wide.

Finding a pelecypod in a sideways (profile) position (Figure 21) the two shells are usually symmetrical and are not on the brachiopod. Not really that confusing: looking at a brachiopod from the front (or back), it is symmetrical, looking at it from the side (profile) it is

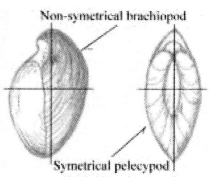

Figure 21 — Profile of a typical brachiopod and pelecypod.

not — and just the opposite for the pelecypod.

Brachiopods were marine invertebrates and remain so today in some deeper sea water environments. The pelecypods, on the other hand, lived both in the sea and in freshwater rivers, and likewise still do today. Clam shells, and the beautifully spiraling gastropod shells, litter the beaches today and are collectively referred to as sea shells. Some clam, or mussel, shells contain the brilliant mother of pearl patina on their interior surfaces. Coastal Native Americans used clam shells for all sorts of daily and ceremonial ornamentation (after they consumed the soft part inside) and were used for trade with interior tribes. Today factories make buttons from them, or we store them in the basement as a forgotten reminder of our trip to the beach.

Plate 50 — The marine clam *Allorisma*, found in a Pottsville shale below the Coalburg Coal. About two inches long.

Another common pelecypod cast is shown in Plate 50, in this case the view is from the "top front," or opening end of the shell, showing the concentric and deeply-striated exoskeleton of the marine clam *Allorisma*. Not seen in the picture is the lower (ventral) shell which is still hidden in the shale. A strange-looking clam to be sure, but quite common in the marine shales of the Kanawha Valley.

The *Allorisma* shown in Plate 50 was found in a rather dark organic shale but is also found in the silty shales which occasionally separate the Winifrede Sandstone, which, of course, means a marine environment had to have existed for short periods of time while interrupting the freshwater delta deposits of the Winifrede Sandstone.

Yet one more example of a 300-million-year-old pelecypod found in the same marine shale member as *Allorisma* is shown in Plate 51. This little bivalve (of the genus *Dozierella*) looks more like an arrowhead than a clam — possessing a straight hinge line and a triangular-shaped shell. I have found this little sea creature only in the aforementioned shale layer associated with *Allorisma*, and haven't found any other reference to it being found anywhere in the Kanawha

Valley. This same shale layer has also produced several type of brachiopods.

The picture shown in Plate 51 shows only one shell of the pelecypod. The other side, still concealed in the rock, looks much like the exposed side. And like all other pelecypods (and brachiopods) this strange-shaped little clam opened and closed along its hinge line — in this case a very straight and pointed hinge line.

During the Pennsylvanian Period there were hundreds of different genera and species of marine and freshwater pelecypods and gastropods, likewise there were hundreds of genera and species of marine brachiopods. The reason there are no pictures of gastropods (snails) is simply because I haven't found any in the marine *or* freshwater shales in the valley, although in Chapter One (Figure 6) an illustration of a gastropod was shown and discussed.

All, or any, gastropod fossils likely to be found in the valley will be coiled and spiraling upward similar to the one shown in Figure 6, the difference will be in the number of coils and the height of the spiral and the shell decor. Some only spiral once or twice and need not be pointed at the top. Likewise, I haven't found any cephalopod fossils (Chapter One, Figure 7) in the valley; they, too, spiral but in a horizontal plane away from the center (they don't spiral upward but spiral outward like a coiled watch spring). There is no doubt these invertebrates are somewhere in the numerous marine shales that outcrop here. Look for gastropods in either marine or freshwater shales (and the marine Ames Limestone) and cephalopods only in marine shales (and the marine Ames Limestone).

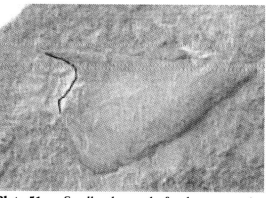

Plate 51 — Small pelecypod of unknown species found in shale layer below Coalburg Coal.

An especially interesting aquatic invertebrate, the majority of which were marine with but a few freshwater species, are the bryozoans (Plate 52). Resembling coral, bryozoans (a catchall name) were very small animals, most less than 0.04 inches (1 mm), which grew together in colonies all the while producing an exoskeleton of calcium carbonate which they absorbed from the surrounding water. As the colonies grew so did the housing structure, branching in a genetically-programmed way, ultimately growing into a large (relatively speaking), fan-shaped, high-rise apartment made up of single room suites.

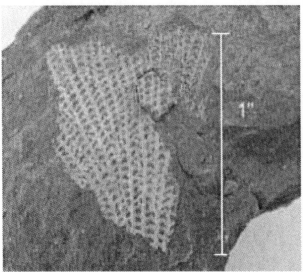

Plate 52 — The bryozoan colony of *Fenestella* grew in an upright, fan-shaped structure of its own making. Found in the Kanawha Black Flint, Pottsville age. About one inch high.

The bryozoan shown in Plate 52 was found when a piece of Kanawha Black Flint was split apart, a remarkable find for its preserved detail and delicate colonial structure. Bryozoans and some coral are considered colonial animals because they grow together in one structure (said to live in colonies); animals like brachiopods are solitary individuals.

Bryozoan colonies commonly grew to an inch or so high and usually attached themselves to the ocean bottom — although some species did get larger (up to two feet across). Several species didn't grow "up" at all but grew flat on the backs of brachiopods and pelecypods, encrusting their shells with calcareous secretions.

Bryozoans are known from the Ordovician to the present and do exceptionally well in a warm, shallow water environment with mild currents, commonly in a limestone or shale depositional environment. Obviously, since this one (Plate 52) was found in the Kanawha Black

Flint, the depositional environment of the flint (when it was still a shale) may be deduced. Their colonies, which in some cases are mistaken for coral (no relation), grew on the sea bottom attached to a rock or some sea shell. The animals themselves actually lived in small, individual, open calcareous tubes from which they extended filaments, or minute, sticky tentacles, to capture drifting plankton.

Plate 53 is a detail of a section of the bryozoan structure shown in Plate 52. Each tiny hole represents the home tube of a single bryozoan. Note that the external skeleton consists of a series of almost parallel, but not quite, vertical "structures" supported by cross-over bars, called diaphragms, with large empty spaces in between. The tiny "speckled" dark spots on the structure are the bryozoan chambers.

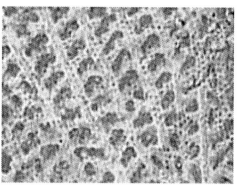

Plate 53 — A close up of *Fenestella* showing the tiny holes, or living chambers, of the individual bryozoans.

This particular high-rise is known as the genus *Fenestella* and, as mentioned above, was found in the Kanawha Black Flint member of the upper Pottsville Series.

Bryozoan fossils are rare in the strata of the Kanawha Valley, one reason is because of the scarcity of marine limestone layers which seem to preserve these animals extremely well. As mentioned in Chapter Two, the Ames Limestone of the Conemaugh Series is thin and not well developed in the valley but may contain some marine fossils locally from, say, South Charleston to Cross Lanes — finding the Ames Limestone in the outcrop is the real struggle. So far I have found the bryozoan only in the Kanawha Black Flint and have not seen any other reference to them.

Even more so than the gastropod or cephalopod, finding a trilobite in the rocks of the Kanawha Valley would be a rare find indeed. One reason is the trilobites were marine animals which had declined in number by the Pennsylvanian Period, thereby reducing the number of potential fossils. Another reason is trilobites lived in a warm, clear water environment where conditions favored sea bottom accumulation of calcium carbonate ooze (a limestone-forming environment). Today they are found most commonly in older limestone layers extending from the Cambrian Period to the Devonian Period during the time when they were most abundant. In spite of this, the marine shales as they occur sporadically in the valley may still hold this elusive critter.

However...the remarkable specimen shown in Plate 54 was indeed found in limestone in the Kanawha Valley, not in outcrop but in a driveway where crushed limestone had been spread.

Of course the limestone was not from the Kanawha Valley and has absolutely nothing to do with the rocks found in our local hills, but is a great example of what may be found no matter where you are. Rocks are everywhere, look at them — you may have a prehistoric creature somewhere around your house too. The trilobite shown in Plate 54 is of the genus *Calymene* which flourished in the warm and widespread seas of the Silurian and Devonian Periods. As old as the rocks are in the Kanawha Valley (and they are old indeed), the trilobite that caused this magnificent cast perished in the lime ooze some 90 million years before the oldest rocks in the valley were deposited.

Plate 54 — A flawless cast of the trilobite *Calymene* found in crushed limestone. Courtesy of Diane Carpenter.

Before I conclude this chapter and leave the subject of animal fossils all together, there is one more that begs to be discussed. Like the trilobites, crinoids (Figure 22) were for millions of years a very successful and prolific family of marine animals. But unlike the trilobites the crinoids are still very much a presence in the shallow sea today. In many Pennsylvanian-age limestones beds they are so numerous they make up the bulk of the limestone itself.

Crinoid is another one of those catchall names for a marine animal that looked more like a plant than an animal but was related to the starfish, sand dollar, and sea urchin, and is commonly referred to as sea-lily. Picture a multi-armed starfish setting upside down atop a long, self-made stalk projecting from its backside so its centrally located mouth would face upward. As mentioned, the crinoids did not go extinct at the end of the Permian and are still with us today.

Crinoids were (and still are) marine animals that lived in relatively shallow, clean water environments. They took calcium carbonate from the sea water and produced a column that lifted its many-armed body from the bottom and closer to the tiny food drifting by with the currents. Rarely growing more than a couple of feet high, although I found one column in the Putnam Hill Limestone in Ohio that was five feet long, crinoids were solitary animals.

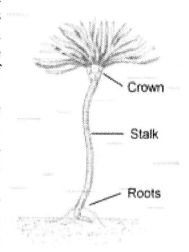

Figure 22 — A typical crinoid (sea-lily) of the Pennsylvania Period.

Sooner or later anyone interested in fossil collecting will come across a fossil crinoid, or at least the crinoid's column parts which are made of calcium carbonate and easily preserved in a limestone depositional environment. The column (stalk or stem) consists of a series of thin disks which look like a stack of buttons (Plate 55) incased in tissue, and ligaments which hold the disks together, but not rigidly, which allows the stem to be flexible and bend with the ocean currents.

Although solitary, crinoids seem to have been very sociable animals, or else prolific reproducers, for when found they are usually found in great numbers. Crinoids also secreted an attachment to the sea bottom, called roots (although they had no "root" function or even penetrated the rocks), which anchored them firmly to one place for life.

By function, crinoids may be divided into three parts (Figure 22): a crown, made up of calcium carbonate plates, arranged differently in different species, which contained the body (mouth, digestive tract, anus) and "arms" of the crinoid; the column, which supported the crinoid; and the roots, which held the animal in place. They fed by capturing plankton with sticky arms which then, with the help of tiny filaments, directed the food toward its centrally-located mouth. Inside

Plate 55 — Segment of a crinoid stalk showing the stack of individual calcium carbonate disks (buttons). About 2 inches long.

the column was a cavity that ran from the crown to the roots, called the axial canal, through which ran a tube of fleshy tissue which also did much of the work holding everything together and adding flexibility.

After the death and collapse of the crinoid, the interior tube of tissue decayed quite rapidly and the cavity filled with the lime ooze from the sea bottom. Interestingly enough, the axial canal (cavity) was shaped differently among many of the crinoid species, some had round cavities which ran the length of the stem, some were star shaped, and some were flower shaped as shown in Plate 56. This is the same specimen seen in Plate 55 but separated at the center crack.

Note that the flower-shaped axial canal is the same color as the surrounding gray limestone. That is because it is filled with the same stuff. Several smaller crinoid stems in various positions are also preserved with this specimen which illustrates the prolific nature of the crinoids in some limestone layers.

Although the crown is not an uncommon fossil in some limestones, the column, or parts thereof, are most commonly preserved and very easy to recognize. After the death and decay of the soft parts, the crinoid (before burial) may break up into the individual "buttons" which make up the crinoid's column and then scattered over the sea bottom.

Free crinoids, crinoids without columns, were once thought to be extinct until around 1860 when thousands of living specimens were dredged from the ocean bottom. Free crinoids were also represented during the Pennsylvanian Period; some grew columns which were then discarded at maturity, some other forms attached themselves to other floating organisms, and some just spent their days sitting on the sea bottom like starfish.

All the different *plant* types discussed in this chapter may be found in the rocks of the Kanawha Valley plus many more. All the *animal* fossils discussed may be found, or have the potential to be found, in the valley plus many more. Often described as snapshots from the past, fossils are Nature's hors d'oeuvres; once you find one it's difficult not to want another. Fossils are much more than a collector's novelty or even a scientific curiosity, they are to the finder a just-opened time capsule...as a matter of fact, every rock layer in the Kanawha Valley is a time capsule with or without fossils. The sand grains tell their story, the bonding agent holding the grains together suggests eons of mineral-carrying ground water, the different types of stone reveal different depositional environments, and unconformities (Chapter Six) show unknown years of erosion. But the organisms that turned to fossils possessed the spark of life, and they carried it well through millions of reproductive cycles. They lived, died, and were discarded in the mud or lime ooze as if they had no meaning. We know, however, that they did have meaning. Every plant and animal existing today owes thanks to these ancient creatures for existing at all. They took the gamble; they came and went until they got it right.

Plate 56 — A beautiful example of a calcified crinoid "button" showing the interior axial canal filled with darker limestone. Found in a lower Allegheny fossiliferous limestone: diameter 3/4 inch.

Chapter Six
Our Heritage

For the apple is still as sweet as can be,
Lord, don't blame me.

Bones, teeth, and lungs; hearts, eyes, and fingernails; what is our heritage? Are we just an upright collection of specialized parts like the stinging jellyfish Man-of-War but with a big brain on top instead of a sail? Is our heritage what we each carry with us every day in mind and body or what we think we know about our ancestors, or both? Did Nature bless us with a loving gift of conscience conscienceness or render unto us a fanciful deception? A Greek philosopher once wrote (in so many words), "Would you rather be a satisfied pig or a dissatisfied Socrates?" These are questions we all wrestle with from time to time and each must be answered according to one's own persuasion. Although we have touched upon life and death processes, and the ancient victories that ultimately lead to the life we see today, in this book our heritage is not necessarily what's inside us, either physically or mentally, our heritage is the ground we live on. Our local landscape is that heritage and includes the Kanawha River, Kanawha Valley, and lest we forget, the Kanawha Hills. We must learn to appreciate our land more than ever because it has literally had its ups and downs; sometimes flat as a board, sometimes under water, and sometimes eroded into rugged hills and valleys — right now we happen to be living in one of the rugged times.

After a heavy spring rain when the Kanawha River turns brown with mud and sediment, it is difficult to deny that this mud and sediment is being washed mainly from the hillsides (it has to come from somewhere). We see the muddy water pouring from the mouths of streams and creeks into the river; we see on occasion the Elk or the Coal River whose watershed has just experienced heavy rains dumping muddy water into the Kanawha. Thousands of tons of sediment that once was hillside being washed into the river and then on to somewhere else. There are only so many tons of piled up land left in the Kanawha Hills and with each rain we can subtract the amount lost; if we only knew the numbers, and given enough time...well, you know.

There's nothing really mysterious about erosion, it always happens when the land is higher than the sea. The only thing water wants to do is reach sea-level (water hates to be up high), so it will keep running downhill until it gets there. The steeper the incline, the faster it goes and the more sediment it can carry with it. Hillside erosion slows to a crawl when rain falls on vegetation but when rain falls on bare soil, either on the hilltops or hillsides, soil and rock fragments are loosened and freed to be carried downhill to the river. As rock fragments tumble and toss their way downslope, they scour away at the stream channel releasing even more sediment into the stream.

Brush fires expose already loose, fragile forest soils to erosion and are also greatly quickened by human activities, activities such as timbering, housing developments, road construction, and even overgrazing. Even the act of water freezing and expanding in the cracks and joints of rocks cause rock particles to be dislodged, pile up, and then be washed away when the ice melts or the next rain comes. The erosion of a hillside is as natural an event as the rain falling on it, but we, the human species, have achieved the status of "King" of the erosional agents — and this is true; water, wind, ice, and notably human activities are now the principle instruments of land form change. I am not implying this is good, bad, or otherwise; it's not a point of view, it's just the way it is. Human activities upset the structural integrity of the topography in the name of improvement while hastening erosion unlike any time in the geologic past, the hills are eroding all right and doing so in plain view of us all.

Look up at the hilltops on one side of the Kanawha River and then look up at the hilltops on the other side — where is everything in between? If you look close at the rocks on each side of the river they seem to "match" up. With but a small dip in elevation, the Kanawha Black Flint outcrops at about the same elevation in the hills on both sides of the river around Charleston but is missing in between. The Pittsburgh Red Shales are missing in between as well until they dip into the valley floor west of Institute and go below the influence of weathering. All the rest are gone too, the Homewood, the Upper Freeport, and the Mahoning Sandstones — all washed away to the great Mississippi delta deposit in the Gulf of Mexico. To get a notion of just how much erosion has occurred during the creation of the Kanawha Valley since the days of the early Teays River system, imagine how much rock and dirt it would take to fill it all back up again.

It seems to have gone on forever; deposition, uplift and erosion, maybe not in this sequence but nevertheless Nature has created for us not only a valley but many beautiful and exceptional erosional features. Many of these have been preserved in parks and other public places; the only requirement is you may have to hike some distances to see many of them.

Plate 57 shows an example of an erosional slice through the Saltsburg Sandstone (Conemaugh) on Little Creek in South Charleston. Here Little Creek has cut a rather sizable gorge in the friable sandstone creating an impressive 30-foot cliff at a point where the top of the sandstone bed is capped by a four-foot layer of harder, more weather-resistant sandstone (visible, top center).

Plate 57 — A ice castle on Little Creek. Here, Joe and John (my oldest son at right), get a close-up look at the impressive frozen waterfall.

The ice display shown in Plate 57 is unequaled in the valley, a splendid example of the local power of small stream erosion and the beauty that so often accompanies it. The ice is also breaking up the rock in places unseen, small cracks and joints are being shoved apart by the expanding ice while thousands of individual sand particles are being dislodged by the ice force. Little Creek is full of liberated sand particles.

While on the subject of erosional features, another impressive example, which happens to be in the Grafton Sandstone (Conemaugh), is the pinnacle called Devils Tea Table (Plate 58). This promontory of resistant sandstone was carved from the valley wall by Trace Fork. Trace Fork has been an exceptionally aggressive stream through the millennia, carving down through 200 feet of Conemaugh-age rock creating the unique and scenic Trace Fork Canyon, a showpiece of steep canyon walls, heavy forests dotted with thickets of pine, ferns, rhododendron, and an occasional *Calamites* descendant. Devils Tea Table can be seen high on the valley wall and is assessable by a trail head in Little Creek Park. Although this isolated pillar is impressive, what's really impressive is what's gone; Devils Tea Table is nothing more than a remnant of a solid massive sandstone formation that once bridged Trace Fork and beyond. Everything you don't see in Plate 58 has been eroded and is still being eroded, especially on the Rt. 119 side of Trace Fork.

Plate 58 — A thirty-foot solitary tower of resistant Grafton Sandstone left unattended by millions of years of erosion. The relatively flat surface of the tower represents an old bedding plane.

Big erosional features and oddities of nature are always fun to experience, especially up close and the bigger the better; that's why millions of tourists, including myself, have gazed weak-kneed into the mile-deep gorge of the Grand Canyon, one of the largest erosional features of all. However, the Colorado River did not cut "down" through the mile of sediment that makes up the canyon walls; instead, the river remained at about the same elevation while the Colorado Plateau raised up through it! A play on words perhaps but one worth noting since this same type erosion and uplifting process is why today the course of the New River seems to magically run north-westward through high mountains at right angles instead of running along beside them like most rivers. As discussed earlier, the New River, or, better yet its predecessor the ancient Teays River, was here long before the current Appalachian Plateau and cut through the mountains as they were slowly shoved up from below.

Excluding human activities in this instance (since we were not there yet), water, wind, and ice took what was deposited in some ancient river channel or marine lagoon and ever so slowly chiseled it into random patterns, leaving some parts stand and letting some parts wash away. Nature's creations are sometimes neat and captivating and sometimes just "ordinary." But whatever the final random result, grand or commonplace, it depends in large part on the rocks themselves, like the type of rocks available (sandstone, limestone, granite, shale, etc.), how well the grains are cemented and cement type (silica, calcite, etc.), whether or not the rocks are fractured or jointed due to earth movements and compaction, thickness of resistant or nonresistant layers, elevation and rainfall, cover vegetation, and on and on.

For the rest of this chapter, the discussion will not be on rock features that are the "biggest this" or "deepest that," although these are extremely important and dynamic natural creations; our discussion instead will be on the small structural features in rocks that the reader can look for in his or her neighborhood or favorite hiking trail and attempt to piece together the history and development of a particular rock layer and also look at how erosion, i.e., water, ice, and people affect exposed rock surfaces.

Every single road cut, exposed cliff, or shallow mountain brook, is a treasure house of prehistoric information that may be appreciated right down to a single quartz particle, and each exposed rock layer is beautiful in its own unique way. In Chapter Two the different rock types found exposed in the Kanawha Valley and the environment in which they were most likely deposited was discussed. With this

WEEPING SANDSTONE

information in mind it is now possible to look up close to the rocks to see what prehistoric events can be deciphered.

As sediment is deposited, as mentioned earlier, it is laid down in horizontal layers unless, of course, the sediment is washing over an incline like the edge of a delta deposit. This is pretty much universally true and it is still going on in the modern world. Sometimes sediment is washed into an area of deposition only periodically, say by cyclic flooding; in the long run this will produced a laminated, or layered, affect in the deposit. With each influx of sediment a new deposit of sand, silt, or mud is deposited onto the previous deposit.

Sometimes these successive deposits are relatively thick but more often than not they are quite thin, varying from a fraction of an inch to three or four inches, and occurring in sandstone as well as shale. In many cases, the bond between each successive layer is weak, even after 300 million years of compaction, and splits apart as bedding planes, as pointed out in Chapter Two when discussing shale.

Most bedding planes occurring in sandstone are more difficult to see and are usually cemented, or bonded, together with calcite or quartz from the ground water which makes their separation difficult if not impossible. Natural weathering of a relatively friable or weak-cemented sandstone sometimes will, however, reveal the individual sediment depositional events. The exposed sandstone section in Plate 59 shows signs of recurring depositional events which means this particular sandstone was deposited in a sequence interrupted by periods of little to no deposition, or, in other words, a freshwater environment with frequent influxes of sand sized sediment. This type of deposition is differentiated from the massive deposits of a delta environment

Plate 59 — An old sandstone cut showing the signs of many depositional events. Several of the more prominent are marked with an "X".

where sediment washed in continuously forming great thicknesses of sandstone with no apparent pause in deposition.

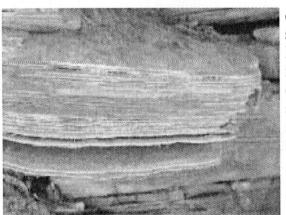

An even better example of a laminated sandstone is seen in Plate 60, here the individual depositional events can be clearly seen and are evenly separating along the bedding planes like a deck of cards.

This particular small ledge of medium-grained sandstone is jutting away from the main body of sandstone situated below, allowing weathering and gravity to cause bedding plane separation. There are around 23 individual layers (laminations) in this exposed sequence (the one-foot section) meaning this spot in prehistoric time experienced some mild periodic flooding of about the same magnitude since the individual layers are all about the same thickness. Each single "slab" (0.25 to 0.5 inches thick) represents a single episode of deposition. The remarkable thing about this is the uniformity and constancy of each small layer and, just like the rings in trees, the number of moderate floods can be counted. If the floods occurred once each year, it took twenty-three years for this one foot deposit to accumulate — but who knows?

Plate 60 — A very thin-bedded sandstone subjected to weathering and splitting along individual bedding (depositional) planes.

Of course there is an alternative to the flooding hypothesis. Maybe the water was there all the time and only periodically charged with sand sized particles washed in as runoff from downpours on higher ground, but if the water was there all the time, one would expect some mud deposition in between the layers of sandstone. Whatever, this one foot section of sandstone was repeatedly interrupted by frequent periods of depositional inactivity; how long the interval lasted between deposition is unknown.

WEEPING SANDSTONE

Plate 60 also demonstrates, in this particular case, the effectiveness of the cementing agent within each layer but lack thereof between layers in this case. This is an exceptional example of frequent depositional events (by whatever manner) accompanied by the randomness and beauty of erosion on a small scale.

Another distinctive feature frequently observed in rather thick exposures of river sandstone is cross or current bedding (Plate 61). Cross bedding can be formed by either water or wind action. It is sometimes referred to as false bedding because they do not represent a true depositional event but instead are formed by directional currents of successive sand deposits over the edge, or tongue, of previous sand deposits, much like the downwind side of a sand dune (Figure 23).

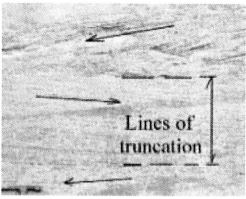

Figure 61 — Cross bedding in the Homewood Sandstone. On fresh surfaces cross bedding usually appears as darker lines caused by the segregation of minerals during deposition. Note the various depositional directions (dashed lines). Horizontal arrows show direction of flow. See Figure 23.

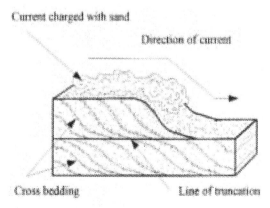

Figure 23 — The formation of cross bedding in a river environment. Charged water sweeps over the leading edge releasing sediment to deposition, all the while extending the "new" deposit further down river.

Cross bedding is usually indicative of a heavily-charged (containing much sediment), fairly rapid-moving river deposit. Note from Figure 23 that the direction of the current may be determined by the inclination of the cross bedding. The odd configuration of the cross bedding in Plate 61 suggests the current, for whatever reason, changed direction for a while as illustrated in the middle of the exposure. Specifically, this type of cross bedding is called herringbone bedding, which does indicates a change in the direction of current. The change may not have been a complete turnaround since the cross bedding in the middle beds are not as "sloped" as the other two extreme beds. How one sees cross bedding, obviously, depends upon the orientation of the cut through the rock layer.

What would happen if the Kanawha Valley, instead of eroding, would suddenly begin to fill up with sediment? This could be caused by a regional subsidence of the land or a rise in sea-level such that the water was now in no great hurry to go anywhere — even during flooding. The hills would continue to erode for a while but the sediment would just wash down to the valley floor and come to rest since the Kanawha River would no longer be free to flow unimpeded through the valley. The river would slowly fill up with sediment and then overflow its banks — the whole valley would then become a very wide, shallow and slow-moving, sediment-filled, sluggish river extending up into each tributary valley — just to deposit more mud and sand along the valley floor where the slow-moving currents would allow. Given enough time, the hills would erode down to the rising valley floor where they would meet in a broad peneplain.

Those of us who think a rise in sea-level (for whatever reason) would only affect cities along the coast may want to ponder this scenario since all water runs downhill to the sea. If the sea was no longer downhill, the water would stay where it is. Right now the Kanawha River pool level is around 570 feet above sea level, or we could say, above the Louisiana and Mississippi Delta, so we don't have to worry yet. After the formation of the peneplain, future geologists could see this event (the filling of the valley with "new" sediment) recorded in the rocks. They could see the old, flat-lying rocks of the Pennsylvanian Period on both sides of the valley with their different layers such as the Kanawha Black Flint now butting up against the new and different rock types that filled the valley. This line of contact between these two different deposits (the old and the new) is called an unconformity simply because the beds on either side of the line don't "match up" — they don't conform. Someone looking at a cross section of the filled-up Kanawha Valley could plainly see that before the new

WEEPING SANDSTONE

deposits were laid down, which subsequently filled up the valley, there must have been a very long period of erosion to create such a big and deep unconformity.

Throughout geologic time there have been random periods of deposition followed by erosion, only to be followed by more deposition, in each case creating an unconformity. Some unconformities are fairly easy to see because of the configuration of the unconformity itself (undulating, sloping, etc.) and others almost invisible, but in each case they represent a period of lost time; that is, the erosional time. One can see how much sand or shale is deposited but can't see how much has been eroded away in an unconformity.

As an example, although on a somewhat smaller scale than the whole Kanawha Valley but an impressive unconformity nonetheless, look at the road-cut exposure shown in Plate 62. The relatively flat-lying sandstone "a" at the bottom of the picture and partially obstructed by brush is overlain by a gray silty shale. Sandstone "a" seems to grade, although rather abruptly, into the overlying silty shale. It may be reasoned that the depositional environment changed from one of moderate water movement to one of more calm water movement hence from sand sized particle deposition to smaller freshwater silt and mud deposition. The bedding planes can be easily seen in the silty shale which indicate a periodic influx of silt and mud mixed with dark organic matter. The silt and mud deposition went on for quite sometime, building up to what is seen today as a relatively thick silty shale layer (about 10 feet on the right side of Plate 62) — but who knows how thick and

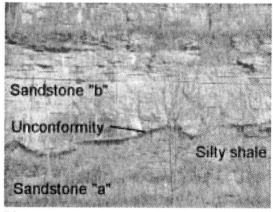

Plate 62 — An exceptional unconformity representing an old erosional surface between a silty shale layer and an overlying sandstone. Unconformities represent lost time intervals where deposition ceased and erosion of unknown extent occurred. This unconformity is in the Winifrede Sandstone - Pottsville Series.

for how long the shale accumulated because the contact line between the top of the silty shale and the base of overlying sandstone "b" is an unconformity.

Somewhere in time, the silty shale stopped being deposited and began to erode — maybe a shift in the direction of a river or a drop in sea level. The shale layer was at the top of the heap when it was being deposited, but the water level dropped relative to the level of deposition, exposing the shale to the weather. Erosion occurred over a long period of time but then stopped (in about the same configuration we see the top of the shale layer today) when sand once again began to wash in over its exposed weathered surface, piling up and forming the massive sandstone "b" exposure now seen in the road cut. Sandstone "b", by the way, is a river sandstone showing much cross bedding on its exposed surface.

Unconformities define intervals of time with nothing to show for it, as in the case of the silty shale in Plate 62. There is no evidence remaining to suggest how thick the shale ultimately got, or if any other rock types were deposited on top of it — all that remains is what's left of the shale and a sort of hammocky or undulating, sloping surface (sloping toward the left in the Plate 62). Had erosion continued until no shale at all remained when sandstone "b" was placed, all that would be seen today is one sandstone formation on top of another sandstone formation with no indication of the existence of the silty shale — or anything else that may have been on top of the shale.

How does the geo-philosopher deal with this lost time interval question? It's one thing to stare at a huge rock outcrop and imagine how long it must have taken to build it, it is another thing to look at an unconformity and wonder where all the days and nights went in between. All that sand was not stacked up at one time though, what we see today is the end product of millions of days, weeks, months, and years of river sand deposition. How in the world do we stop it and concentrate on just one day, if not one hour?

As I reflect on this question, it occurs to me the best way to put the spotlight on a specific event in geologic time is to look at "fossil" ripple marks. Of all the marvels preserved in rocks, the one feature that actually does capture a moment in time is common current ripple marks (Plate 11 and Plate 63). Ripple marks are familiar today along the shallow shoreline of the Kanawha River. As the waves from frequent speedboats or coal barges lap the shoreline, ripple marks are left in the sand momentarily, only to be changed when the next series of waves roll in. There are other things familiar along the river as well, things like discarded cans, old hunks of metal, pieces of concrete and woven cable left over from some turn-of-the-century, coal-loading facility, used bait containers, plastic bags and rags stuck in low-limbed trees, and even old, abandoned barges marooned so long they have become part of the river bank; covered with mud and sand these old barges now have brush and scraggly trees growing from them.

I have always wondered why knapsacks, bags, buckets, and pockets always seem to have enough room to carry full items into a recreational area but never seem to have enough room to carry them back out empty? Am I missing something?

At any rate, finding ripple marks on the surface of a slab of hardened sandstone, one can come to four immediate conclusions: First, the ripple marks were formed by water gently breaking upon the shoreline of an ancient river (ripple marks can also be formed during the final episode of flooding); second, the sand was free flowing; third, since ripple marks frequently change their pattern depending upon the intensity and direction of the current, they represent a relatively short duration of time; and fourth, the direction of the water that formed the ripple marks can be determined.

Ripple marks are formed much like sand dunes. For the most part, the sand is already there as the shoreward currents (wind in the case of dunes) redistribute the sand into long, parallel ridges perpendicular to the current direction. There are three commonly described parts to a ripple mark: the lowest part, characterized by a gently-sloping, concave surface, called the trough (like the low part of a sand dune); the highest point (or part) called the crest; and the leeward side (the sheltered side — the side sheltered from the current), usually sloping rather abruptly into the next trough.

Plate 63 is a remarkably well-preserved example of current ripple marks found in a thin sandstone layer (about four inches thick) some fifty feet below the Winifrede Coal. This sandstone fragment fell from its cliff-side neighborhood, where it had been in repose since the

Pottsville Period, into the roadside ditch below and became just another piece of nondescript rubble. An excellent specimen indeed; from crest to crest each ripple mark measures roughly two and a half inches across and the width of the whole specimen at its widest point measures almost two feet. The diagram at the bottom of Plate 63 shows the profile of a couple of ripple marks and the direction of the current that formed them — the white stuff beside the sandstone slab is snow. What a beautiful day it was when I took this picture.

Plate 63 — Ripple marks in hardened sandstone. Mute testimony to an oscillating, shallow water current one day 300 million years ago. Note, current direction from left to right which means the shoreline was to the right.

There are many different kinds of ripple marks, the type shown in Plate 63 are appropriately called oscillation ripple marks. Others include interference ripple marks (honeycomb appearance), double crested, trough, etc., but the one feature all have in common is that they each represent an instant in time — they had to form during the last gasp of an onshore current. As the water rapidly receded, the last-made ripple marks were left high and dry. Being exposed to the weather, especially rain, would certainly have destroyed them, but instead they dried out and became somewhat hardened, and then were again quite rapidly inundated with sand-carrying currents. Buried before they could be destroyed, the surface of the ripple marks was preserved and became a bedding plain between two different depositional events. Although I referred to these structures as "fossil" ripple marks, they are by definition not really fossils at all but tractive structures, which just means they were produced by currents or waves.

Another erosional feature commonly seen on older sandstone exposures (old road cuts, logging roads, etc.) is a weathering phenomena known as exfoliation (Plate 64). Exfoliation may result from either physical or chemical weathering and consists of the spalling or peeling off of concentric plates from the surface of the rock mass. Probably caused largely by differential stresses (freezing and thawing) within the rock, this type of weathering results in a rounded surface, much like it has been sculpted. We have all seen these rounded rocks, some quite massive protruding out from some old quarry or roadcut. Occurring more frequently on larger exposures of medium hard, yet somewhat friable sandstones, exfoliation does not seem to "cross over" bedding planes as shown in Plate 64; apparently the bedding plain does not transmit individual layer stresses.

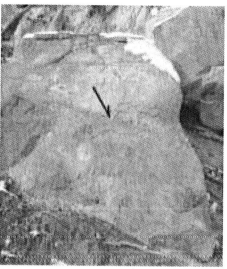

Plate 64 — A dislodged section (about two feet high) of Coalburg Sandstone rounded by the erosional process called exfoliation. The arrow marks the bedding plane.

What undoubtedly started out as a broken-off hunk of angular sandstone (Plate 64) now possesses soft corners and a rounded profile due to exfoliation. Hard, well-cemented sandstone seems to resist this kind of erosion and soft, poorly-cemented sandstone just crumbles before it gets to this stage. Other rock types like limestone and shale seem not to suffer exfoliation simply because they are either too hard or too soft. One could immediately describe the sandstone in Plate 64, just by looking at it, as being weakly cemented.

When next to a rock outcrop, get down close to the contact between two different beds, see if you can determine if the transition between one and the other was slow or abrupt; see if the shale slowly grades into the fine-grained sandstone, see what is directly above a coal seam or directly below. In our world, Nature allows events to happen as they happen and it was no different 300 million years ago. If conditions were right for a swamp, sure enough a swamp developed; if some mud and rock found itself exposed to the weather they eroded, and if a river became obstructed in one direction it just found a new path — back then there was no plan, just too bad for the animals and vegetation that happened to be in harm's way. The difference between now and then isn't Nature, but the presence of humans — now rivers are dredged to keep shipping lanes open, dammed to generate electricity and to prevent downstream flooding, and levied (as on the Mississippi) to create more land to develop; swamps are drained, and hilltops are leveled and then restored in some unnatural design.

We live in the hollow of an eroded peneplain, not because Nature wants us to or even cares; Nature covets nothing, all the while creating and destroying at the same time — when something is lost, something else is gained. It's only when we get in Nature's way do we feel threatened; hurricanes, tornadoes, earthquakes, volcanoes, and floods have been occurring forever and, as already mentioned, no one noticed. No one noticed until we started to build our homes next to big round hills with smoke coming out of their summits, no one noticed until we knowingly built a big city right on top of an active fault line, no one noticed until people decided to build homes in tornado alley, and few people noticed when they settled in river valleys prone to periodic flooding. We live comfortably in this great river channel today because of man-made obstacles built up of rock and concrete put in Nature's way: Summersville Dam on the Gauley and Bluestone Dam on the New.

Previously I alluded to how neat it is to stand back and look at a large rock outcrop or the benched relief of a high road cut, or gaze in disbelief at the mile-high canyon walls of the Grand Canyon. And this is true because at a glance you see millions of years of deposition; you see the many colors which betray the rock's content or cement, shades of red to orange which imply one of several oxides of iron; various shades of gray, calcite or quartz; and dark gray to black, organic content. However, instead of gazing at the rocks from a distance, for the last couple of pages in this chapter I want to zoom in and get closer to the outcrop and look at the actual contact between the different strata and try to decipher the natural events that occurred.

Plate 65 shows roughly the upper five inches of one of the many coal seams found in the Kanawha Valley, but as you know by now there is much more to be seen here than just some coal. This small section shows hundreds of years of sediment deposition and environmental change. First, the coal was deposited as debris falling from countless primitive plants accumulating in a swamp. The plant debris consisted of all available vegetation growing in and around the swamp, like the lycopods, *Calamites*, ferns and seed ferns, and a hodgepodge of anonymous climbing and creeping forest floor vegetation. I wonder what this swamp, found right here in Kanawha County really, looked like, what did it smell like, or sound like on any given day or hour. It was here all right, the proof is in the coal and there are many more seams above it and below it.

Plate 65 — The transition sequence between coal, shale, and siltstone. In this case the transition is relatively "clean," that is, one type grades right into another.

In this spot, swamp (Plate 65) conditions lasted for perhaps many hundreds of years (the compacted coal seam is about ten inches thick overall) and then, ever so slowly at first, the swamp water began to rise relative to the lower-growing swamp plants. As the lower plants were drowned out the deeper, yet passive, water brought with it clay sized particles which settled onto and around the plant debris. At first the clay particle settled at the same time as some of the plant litter, producing a transition carbonaceous clay layer. As the water became deeper, even the larger trees began to die (or better yet, had no place for new spores to germinate). With less and less plant litter available to the water, the clay sized material finally covered the coal seam completely.

This condition lasted long enough for several inches of clay to accumulate on top of the coal (50 years?) then the velocity of the water began to increase such that the clay particles couldn't settle out anymore but silt sized particles were deposited instead. The silt washed in over the clay, accumulating to several inches, after which the water again increased in velocity, this time depositing sand sized particles (not shown in Plate 65). Sitting on top of the siltstone today is about 70 feet of the Coalburg Sandstone (the sand washed in for a long time). And on top of that is the Stockton-Lewiston Coal (the swamp came back), on top of that is the Kanawha Black Flint (the ocean flooded the land), and yet on top of that is 80 feet of Homewood Sandstone, and on top of that it goes on and on. What I have described here is just a small section of the rocks that are stacked up in the Kanawha Valley; the days, months, and years Nature spent here is mind boggling.

If you look close at the separation between the coal and the overlying shale, there appears some sense of slow grading between the two. The coal becomes more clayey until the clay wins out and puts a stop to the swamp conditions altogether. After a period of clay deposition, silt washed in and was deposited with and over the clay. These geologic events can't be seen from rimrocks or overlooks, but they can be seen closeup along a hiking trail.

What Plate 65 illustrates is a relatively ordinary, non-eventful transition of rock types — the exciting part is just being able to see and interpret something that occurred so long ago. It's common for a coal seam to have an underlying clay, called an "underclay" and overlying shale, called a roof shale, which indicates a rise in water level. One may ask the question, "Why does the water level always have to rise after the deposition of coal, why can't it go down instead, or just dry up?" Indeed, the water level can go down (or dry up), but with it goes the swamp. Once uncovered and exposed to air, vegetation will dry out and disintegrate and the whole deposit will now be exposed to erosion — since it is now higher than the water level. The affect of this, at least what we would see today (or not see today), is the disappearance of the coal seam or partial erosion stopped only by the return of the water. Depending upon how fast the water returned over the partially-eroded seam would determine the type rock overlying the seam as we have already discussed.

I'm sure there are many coal beds we don't see today because the water level dropped which allowed the plant litter to erode, decay, and wash somewhere else. What this means is that if a coal seam still exists, then the water must have become deeper such that the coal was preserved, or the water went down and came back up before all the coal was eroded (Plate 66).

Plate 66 shows the upper portion of another coal seam that is immediately overlain by a relatively clean, coarse sandstone. There are no roof shales and there is no apparent transition between the coal and the sandstone — the sandstone, in this case, is said to have good basal contact since it lies directly on top of and in contact with the coal. What happened here 300 million years ago?

Well, first, we know the coal is older than the sandstone (it occurs below the sandstone) and was formed in a dismal swamp (coal only forms in swamps) right there where we see it today; it doesn't matter right now how thick the coal is since the line of contact is the question. Second, the sandstone is a medium-grained, fairly clean, almost white river deposit. We know it's a river deposit because of the very distinct current bedding which gently tappers to the top of the coal, indicating the current direction was from left to right. The current was also of sufficient strength to carry sand sized particles to this spot. Two completely different depositional environments within a fraction of an inch of one another.

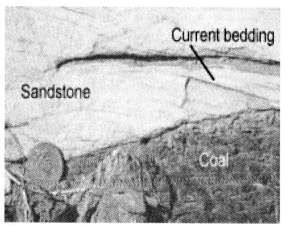

Plate 66 — An unconformity between a coal seam and a freshwater sandstone showing cross bedding. Note the direct contact between the coal and the sandstone.

One possibility for this close encounter would be that a major river draining the eastern highlands suffered a significant shift in direction and immediately began to pour into and flood the swamp, resulting in the sand being deposited directly over the coal. This may be what ultimately happened but not at first. Just barely visible at the top of the coal seam are faint, horizontal compaction planes which appear to be truncated along the dipping (to the left) contact line between the sandstone and coal which are physical evidence of an old erosional surface.

All these clues, that is, rapid sediment transition, truncated compaction planes into the transition zone, and sloping contact line, add up to the conclusion that an unconformity exists between the coal and the sandstone, a lost time interval where the coal, and who knows how thick it was or what may have been on top of it, was eroded to its present surface configuration when sand was deposited due to the coal's sudden submergence. We know clean river sand washed in and covered the undulating and sloping weathered surface of the coal because we can see it. The coal must have been rather well compacted by this time since the bottom of the sandstone contains no coal debris and the coal contains no intermixed sand particles (the coal was already coal, which means the coal is much older than the sandstone; yet the sandstone is sitting right on top of the coal). This all suggests the coal was previously deeply buried and all above subsequently eroded down to the existing coal surface. This sudden submergence of the coal was indeed caused by a rapid shift in the course of an ancient river system. During the Pennsylvanian Period sluggish rivers changed their channels all the time because the land was so flat.

The different rates (slow or rapid) of successive rock type transition illustrated in Plates 65 and 66 each show a coal seam, but it doesn't matter if a coal seam is present or not. Rock type transition takes place during any type of strata deposition or erosion, from shale to silty shale, limestone to sandstone, coal to clay and so forth. The satisfaction is knowing enough about the geological processes to interpret what is there, or isn't there, between two rock types. The excitement is just looking at the rocks with some knowledge of what they are and how they got there; it took a long time to assemble the rocks of the Kanawha Valley and now they are on exhibit as if some huge hands-on, outdoor museum was created for each of us to explore and enjoy. Each rock type represents a different set of prevailing environmental circumstance; sandstone doesn't form in swamps and coal doesn't form in oceans or rivers.

Like other sciences, geology is based on several known physical "truths" or axioms. Euclidian geometry claims that two parallel lines will never meet, or two perpendicular lines will intersect at right angles

(90 degrees) of each other, and on and on. The whole science of mathematics is based on the simple axioms of Euclid. Had Euclid said two parallel lines *will* meet if you look far enough, the science of mathematics would now be based on entirely different principles. Probably the best known edict in geology is Hutton's "The present is the key to the past." This just means the natural forces observed today are the same as those that occurred in the past, and we have no reason to believe otherwise. This being true, we can observe sediment today being selectively deposited by grain size in any active stream or river channel; erosion occurs today from unprotected hillsides, and swamps collect plant debris. Studying the cause and effect of present day sedimentation and erosion allows the interpretation of past sedimentary events. Another rather simple declaration is called, "The Law of Superposition." This means, given two different layers of rock in a vertical sequence, the lower layer is older than the higher layer. As mentioned before, this makes sense since the rocks exposed at the bottom of a road-cut are obviously older than those above — and it's the law.

The earth is alive and well, continually reacting to all the deep, internal natural forces which ultimately are displayed on the surface as active volcanoes and great and small earth movements. As long as these events continue, the earth will stay alive. There is nothing alarming about volcanoes blowing up or spilling lava all over the place unless you happen to live close by — Nature didn't plan it that way. If, for whatever reason, the volcanoes and earthquakes stopped, then we would all have something to worry about.

So, what do we do now, what is our heritage? Well, it's the ancient climatic events still imprisoned in each rock layer; it's the animals and plants that once were alive and now long dead, buried under hundreds of feet of hard rock; and it's the great Teays River that cut our valley out of a flat and pallid plain. Eighteenth-century historians wrote of the beauty of the Great Kanawha Valley and described vast, virgin hardwood forests stretching beyond the furthest mountain like waves on the sea. Just a paltry 250 years ago the Kanawha Valley was the frontier, buffalo drank from the river, Native Americans camped, and Mary Ingles scooped up brine from a salt lick at ancestral Malden...where did the salt come from? Ten thousand years before that a mastodon drank from the clean, cool water of the Teays River while keeping a devoted eye on her young, Native Americans hunted big game, and a natural salt spring released its minerals from deep underground through fissures in the rock which spilled along the shore of the Teays...where did the salt come from? 300 million years before that a cold blooded amphibian stretched out in the sun, a half-mammal, half-reptile-like creature trudged past and left its tracks in the wet sand, and an inland sea was cut off from the vast ocean and slowly evaporated, leaving only its salt content stored between the sand grains of the flush Appalachian Geosyncline...that's where the salt came from.

The salt that rose to the surface in natural brine springs produced the Kanawha Valley's first economy, the timber and coal that followed created much wealth to some of our forefathers and gave jobs to some of our other forefathers. Sandstone was quarried and used to build many of the old public buildings and locks still seen in the valley, and clay was mined for brick and ceramics. All these natural resources were gathered together during the endless Pennsylvanian Period. Ah, the Pennsylvanian Period, with its Pottsville, Allegheny, Conemaugh, and Monongahela Series, still with us today as we drive along the river from the Marmet-Malden area to the St. Albans-Nitro area; you know where you are.

We are living on land we had nothing to do with, but what we've done to it can be seen all around us — from the black-streaked weeping sandstone to the muddy waters, it's Nature's way...can't you see?

Joe on a weathered and thrice-timbered hillside of Pottsville-age rocks on a crisp, winter's day.

Epilogue

Kanawha Valley geology is laid open like a great book, each page representing a different layer with different leading characters. One page stars a million sand grains painted crimson by oxides of iron, on another page fern leaves take center stage. *Lepidodendron, Sigillaria,* and *Calamites* grapple for top honors on many of the freshwater sandstone pages, and on a few pages of dark, organic-rich, marine shale brachiopods perform. All pages are arranged in their proper sequence, each older then the one above. Even page numbers are visible to the enlightened rock reader. No, you won't see page 48 or page 207, what you will see is the Mahoning Sandstone with its petrified *Cordaites* and recognize that it is so many pages above the Kanawha Black Flint; or the very thick page of the Pittsburgh Red Shales and know it is just a few pages below the thin and illusive page of the Ames Limestone.

Rocks are a collection of antiques appreciated for what they are; they need not be cut, polished and sometimes dyed and then eyeballed for their unnatural smoothness and artificial color. What's wrong with just having a hunk of Teays Valley clay setting on the bookshelf, or a generous piece of freshly-broken, white Buffalo sandstone streaked with red hematite setting on the coffee table? And if you collect it yourself (anyone can buy a "pretty" rock) the best part is you know what it is, how old it is, and from what page it was collected.

In this book I have attempted to give the Kanawha Valley rocks their due — every pebble is a repository of prehistoric news, and the fossils they hold are a still life of ancient organisms. A narrative should reflect the sentiment of the writer to which I hope my enthusiasm and love for the subject of geology managed to be unmistakable throughout. But it matters not whether the science is geology, botany, or biology, the gift is the science itself and the joy is in the details. To the reader who took the time to follow closely to the rock sequences as we progressed through the narrative, I sincerely hope you now know where you are geologically — which was my initial objective — but I hope also this introduction to the science of geology has awakened an interest — an interest that will be difficult to shake. The past is waiting.

Selected References

Allman, C. B. *The Life and Times of Lewis Wetzel.* Heritage Books, Inc. reprint 1995.

Billings, Marland P. *Structural Geology,* 2nd ed. Prentice-Hall, Inc., 1960.

Cardwell, Dudley H. "Geologic History of West Virginia." *West Virginia Geological and Economic Survey,* Educational Series ED-10, 1975.

Cross Aureal T. and Schemel, Mart P. "Geology." Geology and Economic Resources of the Ohio River Valley in West Virginia: Part I. *West Virginia Geological Survey,* Vol. XXII, 1956.

Dixon, Dougal. *The Practical Geologist.* Simon & Schuster Inc., 1992.

Dixon, Dougal and Matthews, Rupert. *The Illustrated Encyclopedia of Prehistoric Life.* Smithmark Publishers Inc. New York, 1992.

Dunbar, Carl O. and Rogers, John. *Principles of Stratigraphy,* 4th printing. John Wiley & Sons, Inc., 1963.

Fell, Barry. *America B. C.: Ancient Settlers in the New World.* Wallaby, 1978.

Gillespie, W. H., Clendening, J. A., and Pfefferkorn, H. W. "Plant Fossils of West Virginia". *West Virginia Geological and Economic Survey.* Educational Series ED-3A (1978).

Grimsley, G. P. "Iron Ores, Salt, and Sandstone." *West Virginia Geological Survey,* Vol 4, 1909.

Hale, John P. *Trans-Allegheny Pioneers.* Third Edition (1st ed., 1886; 2nd ed., 1931). Edited by Harold J. Dudley, Th. M. 1971. Published by Roberta Ingles Steele.

Hothem, Lar. *Indian Flints of Ohio.* Hothem House Books, Lancaster, Ohio, 1986.

Hyde, Arnout Jr. *New River: A Photographic Essay*. Cannon Graphics, Inc., 1991

Haught, Oscar L. "Geology of the Charleston Area." *West Virginia Geological Survey,* Bulletin 34, 1968.

Janssen, Raymond E. *Earth Science: A Handbook on the Geology of West Virginia.* West Virginia Department of Education, Charleston, WV, 1964.

Johnson, Patricia G. *The New River Early Settlement.* Walpa Publishing, Blacksburg, Va, 1983.

Krebs, C. E. "Jackson, Mason, and Putnam Counties." *West Virginia Geological Survey,* County Report, and accompanying maps. Wheeling News Litho. Co., Wheeling, WV, (1914).

Krebs, C. E. and Teets, D. D. Jr. "Kanawha County." *West Virginia Geological Survey,* County Report. Wheeling News Litho. Co., Wheeling, WV, (1914).

Krumbein W. C. and Sloss, L. L. *Stratigraphy and Sedimentation*, 2nd ed., W. H. Freeman and Company, 1963.

Lister, Adrian and Bahn, Paul. *Mammoths.* MacMillan, USA, 1994

Lucchi, Franco Ricci. *Sedimentographica: A Photographic Atlas of Sedimentary Structures,* 2nd ed., Columbia University Press, 1995.

McAlester, A. Lee. *The History of Life.* Foundation of Earth Science Series. Prentice Hall, Inc., 1968.

Moore, R. C., Lalicker, C. G., and Fischer, A. G. *Invertebrate Fossils.* McGraw-Hill, 1952.

Peterson, M. S., Rigby, J.K., and Hintze, L. F. *Historical Geology of North America.* Brown Foundation of Earth Science Series. Wm. C. Brown Company Publishers, 1973.

Pirlou, E. C. *After the Ice Age: The Return of Life to Glaciated North America.* The University of Chicago Press, 1991.

Raup, David M. *Extinction: Bad Genes or Good Luck?* W. W. Norton & Company, 1991

Shimer, H. W. and Shrock, R. R. *Index Fossils of North America.* The M. I. T. Press, Eighth Printing, 1965.

Stewart, Wilson N. and Rothwell, Gar W. *Paleobotany and the Evolution of Plants.* 2nd ed. Cambridge University Press, 1993.

Stovall, Willis J., and Brown, Howard E. *The Principles of Historical Geology.* Ginn and Company, 1954.

Tidwell. W. D. *Common Fossil Plants of Western North America*: Second Addition. Smithsonian Institution, 1998.

Twenhofel, W. H. *Principles of Sedimentation,* 2nd ed. McGraw-Hill, 1950.

Printed in the United States
1059200004B/382-384